DOWNTIME ON THE MICROGRID

Infrastructures Series
edited by Geoffrey C. Bowker and Paul N. Edwards

Paul N. Edwards, *A Vast Machine: Computer Models, Climate Data, and the Politics of Global Warming*

Lawrence M. Busch, *Standards: Recipes for Reality*

Lisa Gitelman, ed., *"Raw Data" Is an Oxymoron*

Finn Brunton, *Spam: A Shadow History of the Internet*

Nil Disco and Eda Kranakis, eds., *Cosmopolitan Commons: Sharing Resources and Risks across Borders*

Casper Bruun Jensen and Brit Ross Winthereik, *Monitoring Movements in Development Aid: Recursive Partnerships and Infrastructures*

James Leach and Lee Wilson, eds., *Subversion, Conversion, Development: Cross-Cultural Knowledge Exchange and the Politics of Design*

Olga Kuchinskaya, *The Politics of Invisibility: Public Knowledge about Radiation Health Effects after Chernobyl*

Ashley Carse, *Beyond the Big Ditch: Politics, Ecology, and Infrastructure at the Panama Canal*

Alexander Klose, translated by Charles Marcrum II, *The Container Principle: How a Box Changes the Way We Think*

Eric T. Meyer and Ralph Schroeder, *Knowledge Machines: Digital Transformations of the Sciences and Humanities*

Sebastián Ureta, *Assembling Policy: Transantiago, Human Devices, and the Dream of a World-Class Society*

Geoffrey C. Bowker, Stefan Timmermans, Adele E. Clarke, and Ellen Balka, eds., *Boundary Objects and Beyond: Working with Leigh Star*

Clifford Siskin, *System: The Shaping of Modern Knowledge*

Lawrence Busch, *Knowledge for Sale: The Neoliberal Takeover of Higher Education*

Bill Maurer and Lana Swartz, *Paid: Tales of Dongles, Checks, and Other Money Stuff*

Dietmar Offenhuber, *Waste Is Information: Infrastructure Legibility and Governance*

Katayoun Shafiee, *Machineries of Oil: An Infrastructural History of BP in Iran*

Megan Finn, *Documenting Aftermath: Information Infrastructures in the Wake of Disasters*

Laura Watts, *Energy at the End of the World: An Orkney Islands Saga*

Ann M. Pendleton-Jullian and John Seely Brown, *Design Unbound: Designing for Emergence in a White Water World*, Volume 1: *Designing for Emergence*

Ann M. Pendleton-Jullian and John Seely Brown, *Design Unbound: Designing for Emergence in a White Water World*, Volume 2: *Ecologies of Change*

Jordan Frith, *A Billion Little Pieces: RFID and Infrastructures of Identification*

DOWNTIME ON THE MICROGRID

ARCHITECTURE, ELECTRICITY, AND SMART CITY ISLANDS

MALCOLM MCCULLOUGH

The MIT Press
Cambridge, Massachusetts
London, England

This book was set in Sabon and Avenir by the MIT Press. Printed and bound in the United States of America.

Library of Congress Cataloging-in-Publication Data

Names: McCullough, Malcolm, author.
Title: Downtime on the microgrid: architecture, electricity, and smart city islands / Malcolm McCullough.
Description: Cambridge, MA : MIT Press, [2020] | Series: Infrastructures | Includes bibliographical references and index.
Identifiers: LCCN 2019018828 | ISBN 9780262043519 (hardcover: alk. paper)
Subjects: LCSH: Microgrids (Smart power grids)—Popular works. | Small power
production facilities—Social aspects—Popular works. | Smart cities—Popular works.
Classification: LCC TK1006 .M38 2020 |

10 9 8 7 6 5 4 3 2 1

For Cal, with excess energy

CONTENTS

PREFACE

If you are looking for a bright spot in dark times, please read on. Personally I find a *new grid awareness* downright delightful. You might discover this too. For me, writing this book has restored some daily gratitude and wonder. I feel better about powering things up each day. I now see the city differently. I happily notice electric substations. I can make a bit more sense of the clean technology news. I see past utopians and doomers. I unplug more often, and worry less about technological dependency the rest of the time. When asked what I think of *the smart city*, as alas I so frequently am in my field, I feel like I have something more humane to say. When faced with a sea of tech-forward white papers, all brimming with solutions, I feel a better chance of not drowning in them. If any of that sounds like you too, please read on.

This work has sought to proceed in *science, technology, and society*, a human-centered approach that takes the long view on technological futurism. In this way, I hope to have assembled a wider synthesis, and in that I hope is a more readable package, than industry professionals ever seem to have time to write. If nothing else, this brings the perspective of the university to a fast-changing field. I remain confident that the university still cultivates knowledge in ways complementary to the internet.

Please rest assured that having attempted this, I have become all the more aware of all that I do not know. By attempting such work, so outside my usual domain and so easily able to be dismissed as dotage, I am amazed to have gotten this far. Yet the readers and editors have found something here. For my part, I feel that I have learned something important about living in a *world of systems*, while at the same time learning something specific to all that talk of the smart city. Toward the latter, I hope to have sounded something contemporary for that favorite theme in science, technology, and society, and that most foundational system of them all, electricity.

In all this, I hardly feel alone. By now, so many people want to know so much more about anything clean, green, and local or micro that more than just a few of us deserve to go out and investigate as best we can. You are probably neither an engineer nor a policy analyst, and nor am I. Yet we too can seek a better story. Wherever scholars and journalists can begin to make sense of all those white papers out there, that in itself can help. While there are some good overviews of the grid to be found, alas they also seldom bring enough experiential perspective, nor a focus on the built environment, nor the kind of long-term ethnographic wisdom sought through science, technology, and society. On any of these, I found the academic literature surprisingly scant. Thus I have cast the net quite widely. So in reading this book, please be aware that the work has been developed, framed, and I hope balanced with many potential audiences in mind—for example, in architecture, interaction design, cultural studies of infrastructure, social history of technology, and last but not least, perhaps a few amused readers within the vast and changing electric power industry itself. For when interesting cultural choices lie ahead, sometimes new perspectives can help. However many the narratives from infrastructure builders, policy wonks, doomsday preppers, or green communitarians, still the situation invites new voices. You don't have to second-guess the experts, and this book does not try to do so, to join in this vital new conversation.

Here is how I got to this. After years on the lecture circuit about interactivity in the city, mostly working at the intersection of architecture and interaction design, I do not think I was alone in my dismay. For by the early 2010s, not only had the distant digital overlords taken over and darkened the internet, but also an increasingly stormy world also made it all seem far too fragile. Resilience thus became what to work on, and resilience is mostly local. By now, local means not only food, the arts, and independent small businesses but also clean energy. The 2010s are when local electricity became viable and vital too. The more that the planetary and political climate worsened, the more I found this happy fact impossible to ignore. So when in the year 2016 it appeared that not even a fossil-fuel-sponsored Washington could stop the clean local energy boom, I dared to undertake this study. Perhaps in my own advancing years, I hoped to learn enough to help a local dot org from my public university post. Perhaps somehow I wanted to find a conversation between architects and the big infrastructure builders, who appear more capable than anyone else in this field at the moment. But mostly I just hoped there was yet something good about the smart city.

I began by interviewing dozens of mostly academic counterparts. On this basis, I arranged a month in residence at that epicenter of energy futures sobriety, Lawrence Berkeley National Laboratory. That visit has helped me calibrate much before and since. I continue to read volumes as only a full-time academic has time to do, and with a range quite outside the usual literature. As ever, I seek to write not as a journalist, nor as an expert, nor as a public intellectual, since I am none of those, but instead just in search of some coherent academic retrospect. I hope this result is just that. I trust that a print book is still the best medium for that.

So if you are looking for a more reflective complement to all the latest feeds, posts, and conferences out there, or even if you are simply looking for some thoughts that might last, there is still nothing quite like a carefully edited book. I hope you will find this one quite carefully assembled, and as a result, more readable than all those ones full of analyses and acronyms. I hope this will feel pleasant, unanticipated, and somehow useful. I hope it will hold up as a thoughtful view of a remarkable time. I know it has been worth my while. I do flip a light switch a little bit differently now, and more to the point, I am altogether happier about being here now in such an activated world.

ACKNOWLEDGMENTS

Any book comes about from the contributions of many, and that is worth saying every time. So with apologies for inevitable omissions, let me give thanks of several kinds.

Without question, the first thanks must be to Katie Helke, editor of the Science, Technology, and Society series at the MIT Press, who saw how to shape this work for the Infrastructures series there. Thanks as well to Infrastructures curators Geoff Bowker and Paul Edwards for taking a chance on that list's first work from the field of architecture and urbanism. Also at the MIT Press, thanks to former senior editor Roger Conover for early strategic advice, Justin Kehoe and Deborah Cantor-Adams for seeing the work through production, and designer Yasuyo Iguchi, without whom I cannot imagine doing a book.

Then in some ways, the deepest thanks are to those who offered encouragement at early formative stages of the work: colleagues Geoff Thun and Kathy Velikov, who are seldom without insights on infrastructures themselves; colleague Robert Fishman, who as acting dean at the time understood where my efforts were going; David Nye, master social historian of electrification, without whose encouragement I would have immediately ceased and desisted; fellow technological culture writer and design educator William Braham, whose Architecture and Energy conference at Penn in 2012 was an early indicator of expanding cultural prospects; and Paul Edwards, aforementioned, who while still at Michigan heard some of the earliest notions of this project. The Science, Technology, and Society program here, which Paul's many efforts so sustained, is clearly where this project of mine began. Likewise here at the University of Michigan, I am grateful for the openness of expert colleagues in the Energy Institute, particularly Johanna Mathieu and Ian Hiskins, especially for the general-interest workshop they ran in May 2018.

Thanks to Rick Diamond for enabling my short stay (in September 2016) at Lawrence Berkeley National Laboratory, where despite my lack of scientific credentials I met with receptivity at every turn. Thanks there in particular to Steve Selkowitz for interest in the long view, and to Bruce Nordman for sharing the goal of legible work for a wider audience.

For numerous interviews beyond the above, done mostly by Skype when not face-to-face in metro Detroit or Berkeley, mostly in summer 2016 or spring 2017, and with apologies for omissions, my thanks to Gretchen Bakke, Rich Brown, Lucy Bullivant, Carla Diana, Anna Dyson, Gordon Feller, Harrison Fraker, Harry Giles, Ruari Glynn, Gabriel Harp, Philip Haves, Dan Hill, Rusty Klassen, Carol Menassa, Kiel Moe, Stephanie Ohshita, Janie Page, Chris Payne, James Pierce, Jim Saber, Johannes Schwank, Phoebe Sengers, Elizabeth Shove, Linda Simon, Rebecca Slayton, Jennie Stephens, Margaret Taylor, and Hal Wilhite. I would have soon ceased work without these conversations too. But of course even when not so well prepared myself for quite such a range of greatness, I consistently met with openness. It all made wonder why there are only a few times in university life when quite such a series of conversations can occur.

For generously offering use of images, I am distinctly grateful to the artists João Penalva and Stanza. Without these works, the project would have a different tone. I am also grateful to the many others, mentioned throughout the work, who have furnished illustrations. As almost every query quickly met with a favorable reply, this too has been encouraging.

Last but never least, thanks to my lovelies, Kit and Cal, for hearing out the play-by-play, which on any given day went something like, "Can you believe I am doing an old-fashioned print book, on electricity no less, so far into the twenty-first century?" They did believe it, for despite the times, we do have a house that is not only lined but indeed lived with books.

He stepped back. "The telescope's focused," he said. "Don't move it, just look through." Kirsten looked, but at first she couldn't comprehend what she was seeing. She stepped back. "It isn't possible," she said. "But there it is. Look again." In the distance, pinpricks of light arranged into a grid. There, plainly visible on the side of a hill some miles distant: a town, or a village, whose streets were lit up with electricity.

—Emily St. John Mandel, *Station Eleven* (2014)

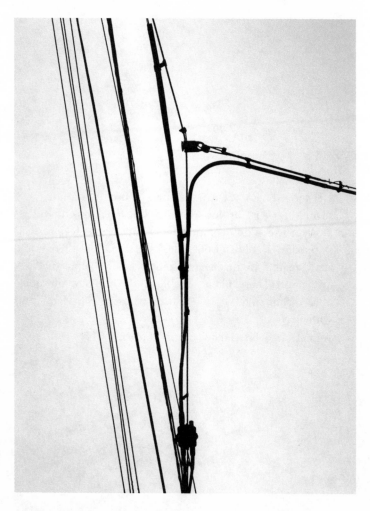

1.1 Restoring some sense of wonder: João Penalva, *Looking Up in Osaka*, Ebisu Higashi cho-me, *Naniiwa-ku* #3 (2005–2006). Courtesy of the artist and Simon Lee Gallery.

1 AT THE EDGE

What would you miss? To live in such a vulnerable world takes some new readiness to let go of *something*.[1]

What would you miss if the power went out? This is not a lament on technological dependency. Those don't get far. This is not a checklist for doomsday preppers, nor a manual for building your own ecovillage, nor a moral guide to greener living. This is not just about installing diesel backup generators. Those spew heavy fumes and soon run out of fuel. This is not a policy analysis, nor a business plan. Despite origins in design disciplines, this is not another vision for a shiny utopian technofuture. Please don't look here for any of those.

If the regional electricity grid went out for a week, what would soon seem most unnatural about that? When did constant electric power start to seem more natural than its absence? How much more of an annoyance would a weeklong outage be to you than it would have been to your grandparents? How much more would there be to miss now than, say, fifty years ago, when as the joke at the time had it, "If it weren't for electricity, we'd all be watching television by candlelight."[2]

Where would you prefer to be in an outage? What kinds of institutions make good meeting points? What kinds of places should plan for resilience? What happens when some have power and some have not? Although critical sites like hospitals, fire stations, or server farms already have their backups, where else should also stay up?

What if your site generates its own? Does making your own electricity foster more thoughtful everyday use? When does at least some monitoring, tuning, and participation feel welcome? Does having it physically present make it more pleasant to keep in mind?

⚡

You might well read this with both pragmatism and wonder. Today almost any individual, organization, or town might generate some of their own electric power, most of the time. Although interconnecting it all gets quite complicated, this simple advantage can no longer be dismissed as naive idealism. Local electricity has quickly become a practical reality. Today a great many players seek a piece of this boom, and millions of owners at least take comfort that it is happening.

So this is about *a new grid awareness*, based in physical surroundings. It is about finding the role of the built environment in the local electricity boom. It is not so much about energy conservation or emergency preparation, which are both necessary but soon tedious topics, as instead about the rediscovery of abundance, expression of cultural change, and participation in local networks. That might seem an unusual invitation for a matter more usually approached through engineering and public policy, but it has become a worthy one. Even just to restore some sense of wonder (figure 1.1), a new grid awareness is worth cultivating.

In particular, this read is an appreciation of inhabitable scale. In an age obsessed with the handheld smartphone and the vast distant network, any scale between those may need rediscovery itself. In the arguments to follow, places and not just individuals frame cultural change. Buildings and districts mediate the scale of local electricity.

This is about innovation and resilience at the edges. As ecologists, for instance, know quite well, more adaptation occurs around the boundaries of different spatial patterns than in their more uniformly consolidated centers. In an unexpected disturbance, a diverse cluster often responds better than a larger monoculture, however well that has been engineered for more expected variations.[3] Today those abstract systemic principles increasingly apply to local clusters of solar panels, battery storage, two-way distributions, network controllers, and local aggregations for trading and balancing. In a world now full of sensate, adaptive systems, even something formerly so uniform as the electric grid may no longer need one size to fit all. Whereas once most grid innovation took place in central power plants, today it has moved out to the nodes—on millions of sites, integrated into other physical structures and systems, with a growing diversity of owners and operators. In a simple name for this complex phenomenon, strategists now call it *grid edge*.[4]

To bring an experiential perspective to these develop-ments, consider *architecture's grid edge*. That is not an expression in everyday use. While unlikely to go viral, it does provide a convenient name for renewed appreciation of surroundings. After all, most electricity gets used in buildings. Always-on power has disappeared into normalcy. Thus to speak of architecture's grid edge invites new design expressions of living with increasingly local, knowable, and better-appreciated electricity. It reflects the many grid edge elements now becoming components of physical surroundings. There it implies cultural value change. It emphasizes that inhabitation matters: most meaningful social change gets framed by familiar material circumstances. After all, the word "architecture," in its larger sense, means any spatial configuration that people are going to have to live with.

In taking this position, it can help to think of architecture as an accumulation of systems. In that, it can help to think of systems as sets of workable material circumstances. For in a material system, the spatial configuration matters as much as the features of any individual components. The desired performance emerges from tuning this system in the context of other systems. For in the built environment, each new technology layers into the surroundings, and somehow remakes them in the process. Electrification did so a century ago, with profound cultural impact, and likewise information technology does so today. The latter is not just new smartphone apps. Far more technology gets built into surroundings than gets carried about in your bag. Not all of that is in smart homes: in buildings, far more diverse layerings of these technologies occur in larger commercial, institutional, industrial, and transportation settings than in so many millions of similar individual dwellings.

This is an invitation to rethink major categories like "the" grid, or what back in the twentieth century they called "the system." To imagine each of these as one total thing has done little good, especially if it ever seems out of control. Instead, you probably sense a growing need to adapt amid a *world of systems*. Within this ever-changing world, some of the largest systems seem admittedly vulnerable, but also more variable, and increasingly able to be interconnected, and intermittent.[5]

So if you encounter hype on *the smart city*, or wonder what, if anything, that means to your work, you might start by thinking of it as many and not one. You might ask when, where, and for whom. If you hear talk of a *smartgrid*, you might look for its innovations at its edges, among the millions of devices now clustering there. To find something real and positive amid so much smart city futurism, consider the physically embedded context of its power systems. There at architecture's grid edge, today anyone might face new priorities. In those, you might expect that the worthwhile opportunities are local, or at least the worthwhile questions are so. If so, read on.

⚡

Opportunities in local electricity futures seem most relevant for the billion people who never had much electricity in the first place. Just as mobile phones have especially transformed places that lacked any previous landline service, likewise local electricity makes all the more difference in places without a reliable electric grid. Now as more local arrangements become viable, development can often bypass the difficulties of getting onto a grid. Although the results seldom match the very real advantages of a larger regional service, particularly at balancing diverse sources and schedules, alas much of humanity never had such a service.

That is not just a technical challenge. Anthropologists who study this social process assert how electric power gets "advocated, contested, ignored, . . . negotiated and selected" differently in one culture than another.[6] As ever, the success of externally imposed infrastructures depends on how well they fit with social practices on the ground. More local programs can have many social and economic advantages, and more attractive political arrangements. Many more protagonists gain the opportunity to build local prosperity without becoming the bottom of someone else's economic pyramid.

Whatever the circumstances, little else provides such a direct benefit. A famous graph (figure 1.2) using United Nations data shows a high correlation between electric power usage (per capita) and Human Development Index, which is still one of the better measures of a better life.[7] At least that is so throughout the middle range of electrical usage, between extremes of deprivation at one end and waste at the other. To do some good today, assist a solar village project someplace where phone minutes are the currency, where charging stations are the new well (the casual everyday meeting place), and where fuel for cooking would otherwise be foraged wood.

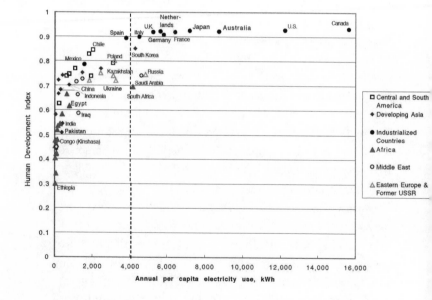

1.2 The famous graph of correlation between electricity consumption and the Human Development Index. Alan Pasternak, Lawrence Livermore National Laboratory, 2000.

However positive and important that faraway story now grows, in its countless instances, meanwhile change comes closer to home. To the billion people in places that have had electricity longest, and among whom the gentle readers here surely reside, new challenges arise. Existing conventions can hardly continue. Likewise here in North America, which is the locus of this writing, interesting choices lie ahead.

⚡

The United States' twentieth-century electric power grid has been called the greatest invention ever. In a retrospect written just after the millennium, the National Academy of Engineering formally declared it number 1 on its list of "20 Greatest [American] Engineering Achievements of the 20th Century." The grid delivers power everywhere, reliably, amid ever-changing circumstances, almost entirely without storage, to be produced and consumed in the same instant. Today, though, this most fundamental technological marvel of modern life is changing more rapidly than it has for a lifetime.

In doing so it becomes a story for anyone. Having long since disappeared into daily life, with its operations invisibly left to engineers and policy analysts, now electricity comes back into the open again and back into more widespread cultural debate. There it invites new forms of appreciation, expression, ownership, and participation. If nobody can agree quite how, almost everyone can agree that major, epochal change has begun. For instance, in 2017, chief executive Anna Pramaggiore of Chicago's Commonwealth Edison observed, "In effect we are turning our own foundation upside down. . . . And this is a revolution. They don't come around often but we are in the midst of one right now."[8]

Now as some historians would be quick to point out, Commonwealth Edison is the very company through which, little more than a century ago, the pioneering tycoon Samuel Insull first built the centralized model of natural monopoly that has prevailed to this day. There began the model that public utility companies still work to defend. So it is noteworthy for such a company to admit upheaval from inside. As this anecdote illustrates from the source, it appears that the question of who gets to play has been cast open.

The aspiring players are many. Whether with voices of economics, environmentalism, urbanism, policy, technology, democracy, or philosophy, many more fields seek new grid awareness. Now as these diverse voices join the conversation, some disclaimer seems necessary. To join this broad agenda does not mean ignorance of long-established expertise. As each discipline explains the scene to itself, it does so with a right to find out, not with better answers. To speak in everyday language does not mean disrespect for the latest complexities in policy. While perhaps no one field can claim highest new insights, many more fields must now seek and express their own, and any one might gain indirect insights from another.

Here in one such grasp of the scene, note four trends, obvious to all, now turning the formerly invisible world of electricity upside down. For brevity, call them energy decentralization, decarbonization, Internet of Things, and external threats.

First, decentralization inherently widens interests. Anyone can now participate in grid reform. As strategists like to put it, "The world is ready to move from energy generated centrally to energy generated everywhere."[9] The cost of wind and solar generation has become competitive with conventional means in many regions (figure 1.3). Local network operations by owners have improved. Storage, the most vital new component, is now advancing more rapidly, under aggressive speculations. Thus the public generally

WHERE SOLAR+STORAGE WORKS SOON

Percent of residential customers with solar+storage cheaper than the average utility revenue per kilowatt-hour

Solar+storage at 11.7¢ in St. Louis

2022 prices

ME	73%
VT	81%
NH	71%
CT	78%
MA	56%
RI	81%
DE	3%
NJ	79%
MD	71%
DC	0%

ILSRs
ENERGY
DEMOCRACY
INITIATIVE

1.3 Electricity generated everywhere: a forecast map of the percent of households by state where power from solar plus storage will cost no more than grid electricity. Institute for Local Self-Reliance.

wants renewable energy and is unimpressed when utility companies say that it cannot yet work.[10] Alas, this bright green democratic future brings dark times for existing utility operators. You don't have to be a policy analyst to see how defectors to locally owned generation leave everyone else with higher per capita rates, which induce still more defectors. And you don't have to be an engineer to see that the variability of wind and solar power, and especially the two-way flow of power, make the challenge of scheduling and balancing it all only worse. In that, the need for long-distance infrastructure does not disappear. Since the greatest diversity of sources provides the best means of balancing ever-changing supplies and demands, the best practices of power quality remain grid-wide. Almost everything about that technology, alas, has been designed for one-way power flow, and much about policy seeks to perpetuate that. Hence these issues have become everyday fare in green journalism.

Second, for larger reasons of climate justice and carbon reduction, many more systems must move to cleaner energy. The cleaner that electricity becomes, the more things should use it instead of using fossil fuels. So long as fuels are being burned, it remains physically more efficient to use the heat and pressure directly, rather than first convert it and then transmit it as electricity. Yet it may not be cleaner. When measured against emissions as opposed to physical efficiency, an equivalent amount of work done with remote electric power can be cleaner than done with so many millions of smaller sites of combustion. A gigawatt hour from a large thermal power plant is much cleaner than twenty years ago, thanks to recent shifts from burning coal to two-cycle natural gas. For the moment, and despite its external problems, that conversion has been North America's largest recent means of decarbonization. Thus while no electric vehicle should claim to be zero emissions until charged from sources that are zero too, already a fleet of them might be cleaner than an equivalent number powered by their own internal combustion engines. Then as generation increasingly comes from emission-free sources, many other uses go electric as well. If ever all sources of electricity were clean, then everything should be electric.[11] Whether such an increase in demand for electricity could drive better grid reform is a more delicate question, however.

Third, and as the main basis for the smart city, digital devices infuse all of this. A culture of information technology that has creatively disrupted resource flows in so many other walks of life now also reshapes the electric power industry. The result is almost always called *smartgrid*. By now that can mean just about anything. But in short, it most usually describes the versatile, real-time, two-way coupling of distributed supply and variable demand. It means new markets, practices, and platforms. As usual for the Internet of Things, it embraces more monitoring, tracing, and tuning. While it has always been important for energy-intensive users like factories, the Internet of Things also now becomes an attractive distinction in commercial designs and a lifestyle element for always-online inhabitants. On the whole, embedded intelligence does tend to increase the viable range of variations. Thus in a 2015 report on the "energy revolution," the business magazine *Forbes* summarized it: "In effect, energy is the last domino of the information age. . . . Clean energy is one of the most dynamic sectors in the world—hot start-ups, technological whizbangery, cutthroat competition, billions in venture-capital investments, a race against the climate clock."[12]

Last but hardly least among these four main driving trends, ever more unpredictable threats arise. For many people this is enough in itself to create grid awareness. Alas, to recite too much of this quickly makes for gloomy reading, but in short, as has become all too evident in the daily news cycle, an aging infrastructure faces increasing disruptions.

Aging makes even its smallest everyday faults more troublesome. In an unequaled tour for the general reader, anthropologist Gretchen Bakke observes how badly the grid has been "fraying."[13] In an annual review for the year 2017, at this writing, the American Society of Civil Engineers gave the entire energy infrastructure a barely passing grade of D+.[14] While most everyday faults come from the usual small sources like a tree falling, the grid's overall age often compounds the effects.

Then the big disasters hit. If nothing else prompts grid reform sooner, it happens right after superstorms. Storms are disasters that anyone might understand. By contrast, although the different external threat of cyberattack is real, it has not yet not been dramatically realized in North America. A successful attack on a large enough substation could bring down a region for weeks, and the cybersecurity to prevent that adds yet another layer of central infrastructure costs. So far the realized threats come from unprecedentedly heavy weather.

Where more than distribution has failed, the outages are not just overnight (figure 1.4). Here already are new haves and have-nots; repairs do not seem to occur equitably. For instance, at this writing, Florida had been impressively restored after Hurricane Irma, but much of Puerto Rico faced months without electric service after Hurricane Maria. As a consequence of its humanitarian crisis, the case of Puerto Rico became newsworthy as a forefront in debates on energy democracy.

1.4 Outages are not just overnight when major equipment fails: a RecX "spare tire" recovery transformer arrives at a substation. Courtesy of ABB.

Who gets to play? Why would anyone who is not an economist, engineer, or policy analyst care to read (or write) about grid edge developments? To the rest of us, it does seem like one of the more positive trends of the times. Throughout the decade of the 2010s in North America, the field has rapidly diversified. Grid reform has been accelerating, and no amount of denial from Washington has been able to stop it. Developments in *clean green local micro* energy have become community causes, legislative debates, investor crazes, and mainstream news.

Of course the investor-owned public utility companies say they have got this. Only they have the means to interconnect, maintain, and operationally balance so much large, dangerous infrastructure. Much of this is indeed an engineering consideration; what happens to stabilize such giant voltage waveforms in the first few seconds and minutes after a disruption involves some impressive physics, and is best done across a large, diverse network. Yet the situation is inevitably also political. The utilities have every incentive to maintain consensus that their monopolies are natural. Yet today they are no longer the only ones providing power. They are no longer necessarily the most innovative builders of energy infrastructure, especially its new layers of information infrastructure. They no longer have the only legitimate positions in energy policy debate.

In a prominent case of what sociologist Michel Callon has described as a *hybrid forum*, today a heterogeneous mix of voices counters the utilities' aging monoculture (figure 1.5).[15] Neighborhoods, governments, dot orgs, industries, commercial real estate developers, individual homeowners, and corporate server farms all weigh in. These many counterparts in this debate do not lack expertise. Major infrastructure builders like Siemens or Schneider Electric propose innovative technological solutions. Online

Infrastructure
Builders

Architects and
Urban Designers

Public
Utilities

Community
Solar Groups

Real Estate
Developers

Commerce
and Industry

Neighborhoods

City
Services

Wealthy 1%

Market
Analysts

State and Local
Governments

Citizens /
Ratepayers

Data
Harvesters

Social and
Environmental
Organizations

1.5   Joining the hybrid forum.

organizations like the Clean Coalition, Smart Electric Power Alliance, or Institute for Local Self-Reliance counter with civic and egalitarian concerns as well as policy advocacy and innovation. Major environmentalist networks like 350.org expose entrenched opposition. Economists debate how to value resilience and not just price efficiency. Urban planners advocate local energy governance, whether for municipal ownership as in Boulder, Colorado, community choice as in California and now Illinois, or design initiatives as in New York's much-watched Reforming Energy Vision. Thus the politics has urban geography. Energy democracy debates arise not only because so few cities control their own power but also since neighborhood- and district-scale developments are such an important locus of cultural identity.

When expertise itself diffuses, however, and as plurality inevitably redistributes trust, not all may go so well. Alas, the internet hardly guarantees a democratic forum. Debate soon fills not only with disinformation from entrenched controlling interests but also with wild misconceptions from the fringes. Advocates with no real background weigh in on complex matters of engineering and policy. The more utopian or dystopian their scenarios, the more quickly those spread.

Lest this writing be too easily misread as another small drop in that flood, do take a moment to understand what is warranted. This writing has no self-interested agenda to promote, nor any firsthand experience at building new infrastructure, nor any claim to refute better-informed expertise. It just asks how so many different players can claim to have separate, proprietary answers on those. It asks the hybrid forum to find fresh perspective on the built environment. In doing so, it points out that inhabitable scale is often both the locus and the subject matter of hybrid forums. Any expertise is in the continuing relevance of surroundings amid the increasing prevalence of interactivity. From that perspective, any proprietary, always-on technological future needs questioning. From many perspectives, locality needs attention. The idea that one place can be different from another has vitality in many fields. So does the idea that infrastructures and localities remake one another. The disciplines of architecture and urbanism somehow understand this. Indeed buildings and cities have been been radically remade by electrification before. Seen in historic perspective, that aspect of locality has useful analogies today, when pervasive information technologies increasingly become a matter of context. As architecture and urbanism know well, there is much more to context than geocoded position. Habitual activities and institutions shape local differences not only in space but also sensibility. The local material context is where social and technical change couple most tightly, and where they interplay as cause and effect. Without such embodiment in habitual contexts, externally conceived programs don't get far.

⚡

In the spirit of hybrid forum, consider what follows as work in *science, technology, and society*. Although the name of that field is often abbreviated as STS, too many acronyms soon make a text unreadable. Most works on electricity are loaded with acronyms. So by contrast, let this writing avoid acronyms altogether. To begin from a single representative word instead, this diverse scholarly field frequently unites on the cultural study of *systems*. Although that work mostly arises from social historians of technology, sociologists of science, or analysts of technology policy, it often takes an interest in the built environment. There it takes an interest in the material circumstances of living with technological systems. There it emphasizes how systems have become so intrinsic to modern life that not everyone senses their history, paradigms, or unintended consequences.

A *system* is a set of processes and things brought together, and operated for the advantages it provides for stability, scale, and dynamic control.[16] Its advantages are mostly internal. Its structure of accumulations, flows, and feedback loops is of greater interest than any one device it contains. This structure makes it behave differently in a given set of circumstances than some other arrangement would do. At least within the conditions for which it has been engineered, a system adapts. According to the consensual definition on Wikipedia, *adaptation* is "the tendency of a system to make internal changes needed to protect itself and keep fulfilling its purpose."[17] Choosing "places to intervene in a system," as the ecologist Donella Meadows famously put it, matters especially for more resilient design. Since "systems surprise us," as Meadows advised, it is better to adapt along with them than to expect command and control in advance.[18]

In a much-admired recent history of such ideas, literary scholar and Re:Enlightenment project founder Clifford Siskin explains how system became a fundamental, especially viable way of shaping knowledge and working in the world. "Even in our best efforts to connect system to history, system has been strangely sublimated into an intellectual issue: an idea that carries and accumulates meanings rather than as an object that works in the world—or doesn't—to produce them." To Siskin, phenomena of use shape a system, more so than an idea, law, or formulation can do. Form, situation, habit, and grasp tend to matter; "we need to engage system as not just an abstract concept or idea but as something materially in the world."[19] Beyond thinking of the world itself as a system, as Isaac Newton so influentially did, today this means operating in a world full of systems.

Of course multiple systems might influence one another. A system maintains itself not only through the interdependency of its own parts but also in reaction to externalities, whether random disturbances or recurrent exchanges with other systems. More remarkable systems exhibit not only internal stability but adaptive response to context too. Self-reorganization occurs more easily in clusters. Many emergent advantages cannot easily be formulated, simulated, owned, or governed. Complexity instead invites discovery, tuning, and adaptation.

Such a situational perspective on systems helps explain so much outside interest in the likes of electric power networks. The externalities are not all technical or environmental. Organizational, social, and political context can create disturbances too. Technological systems, especially the largest ones, not only create but also are created by shifting socio-political outlooks. They are not culturally value neutral.

As eminent works in science, technology, and society have often emphasized, electric power grids are cultural systems. Perhaps because the internet has become the main disruptor, too little such critique has addressed the built environment. Back when it was being called "cyberspace," the internet was imagined to negate geometry and place. Today's soaring city rents increasingly suggest otherwise. Today the internet layers onto the city and purportedly makes it smart. On this, science, technology, and society work has frequently admired the role of architecture and the city, but has too seldom come from architects and urban designers themselves. Of course, the built environment embodies and alters attitudes; new infrastructures more often layer onto than outright replace older ones; and together, social and technological change play out amid the contexts of everyday life.

Otherwise, because infrastructure builders too easily assume all is technology driven, and because the distant digital overlords increasingly own it all from their cloud, the latest futurism tends toward the macro scale. As the media critic Ian Bogost has put it, "Everyone is already living inside one big computer."[20] There the prevailing narrative too easily assumes convergence, efficiency, omnivision, monetization, technoutopia, and much else wrapped up in the businesslike language of *the smart city*.

So the time is right for new chapters in the social history of electrical technology, written from the edges, amid changing values, enacted in the built environment and implemented at local scale. Consider this work as one of what must become many diverse voices in that movement.

⚡

Today there is a sticky new word to help frame such conversations: *microgrids*! Whatever it may or may not portend, the word has caught on and propagated. Relative to more cumbersome policy jargon (and acronyms) of the energy industry, this one is viral enough to be considered a meme. When a word catches on, and fresh media feeds take it all over the place, all that a long print read such as this one can do is to offer some retrospect. At this writing, across the years 2016 to 2018, microgrid implementations have been few (around two thousand in America, according to one respected tracker), but the forecasts and discussions have been many.[21] The number of newsletter stories has steadily increased.[22] From many of the larger perspectives just described, these are the years in which the microgrid became a significant cultural theme. On a practical level, it has gained traction from great opportunities in "solar plus storage"—a way to overcome the limitation that the sun goes down each night. It has also become a regular feature of energy futures conferences. More abstractly, it eventually implies much else about local resilience.

In short, *microgrid* means a local energy network that can intermittently run independently of the larger interconnecting grid. To do so, it needs not only local generation but local control too. With a good mix of energy sources, adequate technologies of switching, monitoring, and voltage regulation, plus preferably some battery storage, a microgrid might run independently for any amount of time, in what is called *island* mode. So then its capacity to go in and out of that mode matters most. Thus to emphasize the importance of this dynamic, the noun has been verbed: *to island*. Microgrids can island.

To build a microgrid adds new layers of costs, responsibilities, physical wiring, and technical challenges to a resource that one could formerly take for granted. For this simple reason not many microgrids exist. Since it also adds new layers of resilience, ownership, and appreciation, however, many more are in the works. Those are not to be underestimated amid such precarious times, nor amid the rise of clean local energy. Hence despite the additional cost overhead, and with energy independence in mind, microgrid prototypes now get debated, funded, and built. The idea has certainly become a part of any larger grid reform visions (figure 1.6).

When a period of open *interpretative flexibility* supplants closed old technological certainties, new concepts invite unconventional investigations. So if the idea of a microgrid has a cool factor that exceeds its immediate advantages, it is well warranted at the moment. The word does appear to appeal to the hybrid forum; at some point strategists, utopians, doomers, engineers, economists, policy wonks, citizens, political lobbyists, and community activists each have used it. When so many separate perspectives simultaneously perceive one same phenomenon, and especially when many distinct disciplines each find themselves at work on one same challenge, often a new sensibility arises.[23] It is time for some cultural research into what this meme means.[24]

For instance, even the US Department of Energy acknowledges this development, even when otherwise politically forbidden to mention much else about life beyond fossil fuels. In a set of public education pages about technology development, the department explains, "Because they are able to operate while the main grid is down, microgrids can strengthen grid resilience and help mitigate grid disturbances as well as function as a grid resource for faster system response and recovery."[25]

Microgrid is not a new idea: the century-old district steam plant that predates the mainstream grid is a microgrid of a sort. Microgrid is not a yet a major market trend:

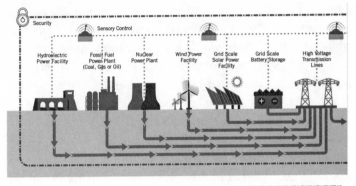

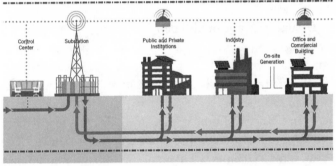

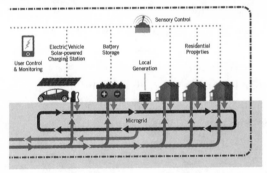

1.6 Microgrid (highlighted here in the bottom panel) in the context of "The Power Industry of the Future," as illustrated for New York's widely watched Reforming Energy Vision (2015). Courtesy of New York Power Authority.

growth is high but volumes are as yet insignificant. So far, microgrids are costly and few. It is not yet a major economic shift the way wind and solar generation have become, although local control of solar plus storage would certainly kick that boom to a much-higher, incontestably advantageous level. A microgrid is not necessarily even green: many run on dirty diesel backup generators. Nevertheless, it has become a provocative way to assert the comfort of grid awareness and the value of resilience. It provides a useful local counterpoint to all those all-connecting, all-seeing, imaginaries of *the smart city*.

Despite the use of the definite article, "the" grid has always been an aggregate; it has always had many distinct regions and pools. Back before they all interconnected, and especially back when each application (streetcars, for instance) had its own power systems, islands of service were the norm. Using them as a part of a larger system had no need for a trendy verb. Today that aggregate quality gets rediscovered. When enough small pieces can intermittently connect or disconnect as necessary, then in the aggregate they can conceivably improve, not worsen, the stability and agility of the larger grid that loosely joins them. Clusters thus can increase resilience.

Ultimately, if such an archipelago becomes indistinguishable from any larger, more centralized entity, it may no longer make sense to speak of "the" grid with the definite article in the singular. It also may no longer make sense to plan so many grids from the top, nor to expect any one technological layer to feed all of them. Thus it was at a microgrid conference in 2017 where Commonwealth Edison's chief executive gave the remarks cited earlier on a "revolution," now "turning the industry upside down," in which microgrids "will emerge as a defining infrastructure."[26]

⚡

As the disciplines of architecture and urban design understand best, a city is an aggregate and not a continuum. In the long read to follow, let the narrative slowly arc toward that prospect. When faced with more totalizing visions of *the smart city*, architecture and urbanism could do more to emphasize this reality. To put it in systems thinking, the aggregate city provides a vital balance between autonomy and connectivity, where specialized zones develop in clusters. Or to put it as a city council might do, sets of premium infrastructure districts have become urban economic generators.

In a foremost principle of urban design, persistent aggregations of built form provide good social infrastructure. Rents soar highest in city centers that would be impossible to imagine without their architecture. Great halls, courtyards, town houses, towers, canal frontages, plazas, and landmarks persist, even as their uses change. Their enduring affordances for good city life surpass the sameness of suburban sprawl or the privations of small town life. Their effectiveness depends on how their boundaries and networks get clustered and nested. There the passage of persons, goods, resource flows, and data is carefully filtered, often intermittently. Then when conditions change, so do those boundaries and flows. Despite the usual use of the definite article, "the" smart city is never one all-connecting system.

At the moment, a search on "smart city islands" brings up a few infrastructure plans for geographic islands (land surrounded by water, such as Chongming's Dongtan Eco-City in Shanghai), and not for islands within infrastructure itself. There is less to be found about capacity for intermittency surrounded by requirements for connectivity. Turn the phase around, however, and the search results are null at this writing. Despite all the millions of posts about all things conceivably being smart at any conceivable scale, at this writing there is not a one with the phrase "islands in the smart city."

This is surprising because so many prototypes of the smart city have been built in specially planned development districts quite separate from the complexities of the metropolitan core. Like Google/Alphabet's oft-cited Toronto waterfront experiment by Sidewalk Labs, the autonomous vehicle campus in Michigan, or Korea's larger, earlier, oft-criticized experiment at New Songdo, prototypes of the smart city prefer to work from the ground up in their quest for whole systems integration at district scale. In a pattern consistent with the largest urban commercial development projects like Canary Wharf, Hudson Yards, or any number of phases of Shenzhen, these developments do create islands. While their flows of information and capital may be continuous, their engagement with the surrounding city is not. This tempts architects and urbanists to consider the inverse: What would it take for a district to remain very much part of the city, yet be somehow occasionally self-sufficient?

To island one resource temporarily does not mean to disconnect from all others at once. A neighborhood temporarily on an electrical microgrid probably does not suspend water or internet connections as well. To island a resource formerly considered so uniform and universal as electricity raises political tensions too. It creates a new kind of haves and have-nots. Nevertheless, to island responsibly, intermittently, and as a part of a diverse cluster can bring advantages to all.

Thus for islands, in the plural, as a noun, instead the better word has been *archipelago*. It describes not just a plurality but also a meaningful set. Islands in an archipelago frequently share some economic, political, or cultural identity. This implies sufficient density among necessarily separate entities. It implies emergence, but without the possibility of a continuum. In its most usual, geographic use, the word describes an island chain. In architecture and the city, however, it depicts the necessity of contrasts, distinctions, and boundaries within the cluster. Yet almost no version of the smart city mentions archipelagos. You don't have to fear the totalitarian potentials of perpetual connectivity to see, just from that simple omission, that there is something quite wrong with those always-on futures.

Islands interconnect intermittently. Often, in the relations of more loosely clustered small systems with more tightly engineered larger systems, the more interesting opportunities are out at the edges.

⚡

With grid edge in mind, this is the shape of the argument to follow. To declare such a road map in advance may help in examining a complex field of increasingly general interest. Toward that aim, several keywords have been introduced here in advance (figure 1.7). Then in the respective chapters ahead, and in the spirit of new grid awareness, begin with a longer historic perspective on electrification. Given the existing abundance of stories on origins with Thomas Edison and Nikola Tesla, let this one instead focus on situations in architecture and the city. Next consider the overall contemporary challenge of local electricity, as so quickly outlined above, in a bit more detail. Do so with the belief that many more voices now belong in a hybrid forum. When everyone knows things must change, but nobody knows quite how, much more widespread grid awareness seems wise. Then to take the local perspective, take a look at the fast-moving microgrid meme. Do so in an integrative way with hindsight, which is what long-form print does best. From that, consider architecture's grid edge, and approach it more as a manifestation of local electricity than by the more usual agendas of smart green building. Next, given that nothing

Grid Edge	*Millions of on-site, mostly owner-operated elements that now dominate electric power innovations.*
Architecture's Grid Edge	*Grid edge elements experienced as a part of physical surroundings, and performing as components of smart green building.*
World of Systems	*The circumstances of living adaptively amid many systems, not just subject to command and control by "the system."*
Hybrid Forum	*The social and political context for debates on policy change, urban design, and heterogeneous engineering.*
The Smart City	*A techo-utopian quest, whose cultural image always lies just beyond actual developments in informatics for city services.*
Smartgrid	*The project, mostly by infrastructure builders, to create an end-to-end, realtime, two-way, transaction network for electricity.*
Archipelago	*A dense, diverse set of islands with intermittent connections, emergent effects and collective identity.*
Downtime	*Here, not just involuntary outages, but also occasional voluntary intermittency, as well as naturally passive operations.*

1.7   A few key terms in the narrative.

should be called smart without appropriate opportunities for participation, note the importance of interactivity. Microgrids run on a new layer of local digital systems. Together, let this series of investigations begin to inform that larger question of archipelagos.

Since so much writing on grid reform is laden with pitches and acronyms, at least something written for a wider readership might try to tell it otherwise. Here, let this narrative seek a long view and some alternative storylines. Let its language be rich but readable, and do not let its lack of practical acronyms imply disregard for specialists in technology and policy. Let it provide a retrospective portrait of a moment (the years 2016–2018) more than any sort of prediction or plan. More specifically, understand this project as a cultural look at how the microgrid meme has catalyzed not only a new grid awareness but perhaps also the beginnings of a major cultural shift away from taking electricity for granted.

"What would you miss" need not anticipate crisis, however. It can also describe everyday islands. The point is that connecting everything always everywhere may not sustain. Since each island in an archipelago might make different choices, with different local resources, and different physical context, and none would have always everything, this too becomes a fair meaning of the question. It also recasts the issue, however, from a question of technological dependency to a question technological appropriateness. Where is intermittency appropriate? Under the right conditions, might more kinds of natural, passive, and unconditioned operations feel right? This is asked amid a rise in digital temperance. As in recent cultural corrections against internet overconsumption, so with electrical use: might a new awareness welcome momentary, daily, and seasonal patterns of voluntary intermittency? What does it mean to consider that prospect in the context of architecture, where for instance traditional or newly net-zero systems need no additional power, and where, don't forget, at least some physical advantages of the built environment still function passably in power outages. So that becomes a third meaning of the question.

With all that in mind, consider one last keyword: *downtime*. Let that mean more besides outages. Let it hold three different meanings to the question, "What would you miss?" Within the context of electricity, consider downtime as outright emergencies, downtime as temporary islands from larger infrastructures, and downtime as natural operations without need for external power. The word obviously has other meanings too, whether in measuring the high reliability of networks or expressing the high sensitivity of participation. In that sense, people themselves need some downtime. From that, take a physiological metaphor for the limits of always-on systems—one worth repeating in some arguments to follow: a heart must always beat, but a brain needs some rest.

Downtime on the microgrid invokes a rich range of perspectives on a matter too often taken as distantly objective engineering and policy. It implies something about where you are. It asks how acceptably that place might disconnect. In perhaps its most obvious interpretation, downtime on the microgrid describes your island staying up while the larger grid is down. After all, resilience is almost always local.

Here then begins a long read on the microgrid meme, from the perspective of the built environment, with an emphasis on longer shifts in sensibility. Although this writing must seem academic in character, nevertheless it invites the general reader. Although it comes from a design domain, nevertheless it may be best read across disciplinary boundaries, even by full-time participants in the building of a brighter, greener future, even in the electric power industry. Even for them, and now too for the rest of us, local electricity now invites transdisciplinary inquiries, increasingly hybrid forums, and at least a few more kinds of guides.

2 ELECTRIFICATION'S ERAS

"Brothers and sisters, I want to tell you this. The greatest thing on earth is to have the love of God in your heart, and the next greatest thing is to have electricity in your house."[1] This eloquent testimony by a Tennessee farmer in 1940 shows just how recently electricity became universal across America. (The year 1940 may not seem recent amid today's obsession with the now, but in any longer historic perspective on civilization, it is recent.) In 1933, when Congress passed the National Industrial Recovery Act in response to the Great Depression, approximately two-thirds of the nation's households had electric power. In the cities and towns, where the power plants were, and where the novelty of broadcast radio now prompted many more families to wire their homes, this coverage rate had doubled in little more than a decade. Yet of the nation's approximately six millions farms, five million remained "entirely without electric service."[2] In an especially controversial component of the Recovery Act, the Tennessee Valley Authority brought the federal government into the electric power industry. This began a series of major public works, agencies, and regulations that would shape an era. Among those agencies, and founded in 1935, the Rural Electrification Administration connected

2.1 "What One Kilowatt Hour Means to the Farm": a diagram from the Rural Electrification Administration.

half the nation's rural households by 1950 (figure 2.1), and almost all by 1970.[3] In 1983, the publication of "The Next Greatest Thing" by The National Rural Electric Cooperative Association commemorated the near completion of the REA's work.

Today that half century from 1933 to 1983 now appears in retrospect itself. Although the ideas that shaped it began much earlier, and the assumptions that it embodied still prevail in many ways today, that era stands in contrast to different realities before and after, not only for electricity, but perhaps also for high modernity. Throughout the late twentieth century, centrally planned world-making agendas increasingly met with unintended consequences, limits to growth, and particularly important for electricity, reversals in economies of scale. For electricity, times both before and after that era involved more choices for more people, and thus more grid awareness.

Before the 1930s, earlier electrification had more diversity of social practices, whether in attitudes, implementations, or adoption rates. Much about that still makes good storytelling, and provides useful analogies about other, more recent waves of technologies, networks, and invisibly embedded activation. Then as the most successful technologies do, electric power slowly disappeared into everyday life. Across the course of that high modern half century, electricity become normal, invisible, reliably provided by others, in the hands of expert authorities, and of no further consideration by everyone else. Whether in engineering, policy, or finance, those authorities fundamentally agreed on how it all worked.

Starting in the early 1980s, however, conditions started to shift, and in the decades since, these have accelerated into a set of changes that nobody can afford to ignore: conservation since the late 1970s; deregulated markets since the 1980s; online exchanges since the 1990s; cybersecurity since the millennium; the idea of smartgrid in the 2000s; a new carbon economics that even the oil companies acknowledge; and under an accelerating and maturing range of these and other trends—viable wind and solar generation to name just one of them—disruptive change in the 2010s.

In 1977, newly elected President Jimmy Carter dared to suggest putting on a sweater if you felt cold. Far less visible in 1977, but with ideas that have become normative today, young iconoclast Amory Lovins published *Soft Energy Paths*, an important early vision of green energy futures. In 1978, Congress passed the Public Utility Regulatory Policies Act, which historians now regard as the end of an era in electrification. In 1982, the Bonneville Power Administration, a showpiece of the New Deal, went into fiscal default from nuclear power boondoggles on the Columbia River. In 1983, economist Paul Joskow published *Markets for Power*, an influential declaration of neoliberal deregulation.

That same year, the historian Thomas P. Hughes published *Networks of Power*, a comparative social history of electrification in Chicago, London, and Berlin, and it soon became a cornerstone of science, technology, and society studies. Hughes was among the first historians to assert that large technical systems are indeed cultural artifacts, and their social histories are cultural destinies. Each of these places in his study "had the same technological pool to draw on, but because the geographical, cultural, managerial, engineering, and entrepreneurial character of the regions differed, the power systems were appropriately varied as well." With a perspective that may seem normal now but was provocative then, Hughes read these cultural differences as a "technological style" to be understood through the "history of systems." As Hughes put it, "A 'System' then means interacting components of different kinds, such as the technical and the institutional, as well as different values; such a system is neither centrally controlled nor directed toward a clearly defined goal."[4] This was a remarkable observation for its time.

Halfway (so far) through this unfolding era, in 1998, the physicist Walt Patterson published what many industry experts came to view as the first widely read, "highly placed heretical book" in their circles, *Transforming Electricity*.[5] Patterson distinctly characterized an era of "destabilization, deregulation, decentralization," which had already begun taking shape by then. "Electricity systems may be the most spectacularly successful technology of the 20th century. They work so well that those who most rely on them hardly notice them," he opened. "Electricity is just there. And so long as it is there, how it gets there is no concern of yours." Now instead, as Patterson freshly argued in 1998, and has become only more true ever since, for ever more people in ever more fields, at work on ever more heterogeneous systems, "electricity becomes something to notice."[6]

Reopening the Imagination

Today a disruptive flood of political, technological, and environmental change demands a much more widely held grid awareness. Today many more disciplines seek their respective ways into this challenge; together they probe their respective literatures and debates, toward new insights on the social history and futures of electrification. Architecture and urbanism can play fresh roles among these disciplines, as they increasingly contemplate a return to local systems, more usable designs of access to infrastructures, and a perennial love of bright city lights.

Thus before jumping into contemporary issues of the built environment, it helps to sketch some long-term background. Historians of electrification seem to agree on three distinct eras (figure 2.2), which to name each in a single word, could be called "innovation," "consolidation," and "diversification." Naturally the events are many, trends grow for decades, and any demarcations by individual dates are somewhat arbitrary. While events seldom align in clear trajectories, nor play out the same in different cultures, historians nevertheless generally agree that at least in the case of America, electrification had, in round numbers, a half century of pioneering implementation, then a little more than half a century of consolidation, standardization, and universal reach, and then already almost half a century (more like forty years so far) of conditions far more heterogeneous and disruptive. Although the transitions may have been

before any grids

electricity a "wild fact"

1880

INNOVATION

early electrification
metropolitan grids
new building types

social experimentation

technological novelties

1930

CONSOLIDATION

economies of scale
interconnected pools
gigantic constructions

social programs

technological determinism

1980

DIVERSIFICATION

deregulation, conservation
computers everywhere
global urbanization

social neoliberalism

technological markets

2030?

after "the" grid

electricity generated everywhere

specialty districts and services

2.2　A time line of electrification's eras.

gradual, and the pivotal events have been many, it seems fair enough and not too much of an oversimplification to describe three eras: before, under, and after "the" grid.

Today almost anyone might expect their company or community to face choices in diversification. Prospects improve in distributed wind and solar generation, local storage, online virtual exchanges and aggregators, and island operations, with smart devices throughout. Almost everyone must admit problems: aging infrastructure decay, increasing risk of cyberattacks, more frequent superstorms, entrenched reactionary politics, and defecting customers. When almost nobody can agree to policies, nor challenge the existing monopolies, nor time the market on pivotal new technologies such as batteries, then much confusion arises—and much more widely held grid awareness seems necessary. Not since the principles and practices of the centralized grid started to converge just about a century ago has there been such change in this fundamental resource.

Let any reader who believes in a bright green technofuture dare to imagine a fourth era ahead, where all things smart, clean, local, and micro eventually totally replace any central grid. Yet let any reader who wants plans and predictions look elsewhere; no one particular futurism is on offer here in this project. There are enough competing prospects to ponder in the present. Does "grid modernization" simply resurface twentieth-century infrastructure, so to speak, without letting go of twentieth-century belief systems? Or does responsive enough become different enough, as the grid becomes the biggest smart thing of all?[7] Or amid neglect of the universal infrastructure might a rush toward local resilience create a new social category of haves and have-nots? Let any reader who believes in a dark near-future feudalism look elsewhere for how to prepare. Neither utopia nor dystopia lies within the scope of this project.

Here instead, ask how to rediscover the scale of the built environment. Explore how that becomes activated in ever more ways yet also provides passive performance, and is universally powered yet locally adaptive. Today as so many things smart, clean, micro, and distributed now remake the central electric grid, some historic perspective can help. Electricity that had quite successfully disappeared into everyday life now invites new kinds of speculation, wonder, and appreciation again. Given how much of the current turn seems toward the local, and given how many of electrification's earliest patterns long ago were urban, surely some such perspective should be seen through the lens of the built environment. Given that no field can afford to ignore the scale of change, given how much electricity gets put to use in the built environment, and given how closely the built environment reflects changing social practices, it seems unwise to leave so many aging assumptions unexamined.

For better perspective, look further back. Today's unfolding era, with its flood of changes, is inherently more difficult to put in historic perspective itself. Meanwhile, the waning of the middle era of electrical consolidation has provided so much of the prevailing reality, for so long, and still has owners who would perpetuate it, that there is less to be gained from reexploring that now. By contrast, to recall that more distant, localized era of early electrification offers freshly distinct lessons today. Let that be the main basis of historic sampling here, and let architecture and urbanism often provide a lens on that.

Innovations, Bright Lights, Big City

Little more than a hundred years ago, less than a third of Americans had electricity in their house. Early electrification spanned a nominal half century, approximately from, say,

1876, when Edison set up shop in Menlo Park, until 1929, when the stock market crash destroyed so many utility holding companies. Historians' narratives still often bring alive this era, when the variety and flexibility of interpretations were greatest, and where the novelty was still distinct. Where public places got lit up on a grand scale electrically at night, it still was a source of wonder. Everyday applications did exist, like taking the streetcar to the brightly lit department store, but each of these would have its own power plant. Most factories still ran on steam, but many were now converting. Steam boilers already in use for mechanical power and heating now also drove local electrical dynamos. Electrically powered assembly lines, which quickly became symbols of the era, could now convey materials to workers and power to specialized handheld tools, far away from any heat and smoke of a furnace, all in well-lit, artificially ventilated space.[8] Electrically powered transit could take trains underground, in subways, downtown to where commuters then rode vertically, in electric elevators, into a new architectural type called the "skyscraper."

A vast and delightful literature exists (figure 2.3). Stories of electrification tend to focus on the earlier stages, when the wonder was greater, and the shape of the systems had yet to be agreed on. Tales of Edison and Tesla nearly support a publishing genre in themselves. Tesla was eminently practical in developing the induction motors and transformers that still drive the electric power industry. Edison's Pearl Street power station opened in New York City in 1882, and is considered the first central power station as well as the first cogeneration in what might today be called a microgrid. Down the years, Edison's light bulb became the icon of having an idea; Tesla's theatrical persona became the icon of the mad scientist.[9] Early film classics surround the mad scientist with dynamos, big throw switches, and lurid neon.

THEME IN EARLY ELECTRIFCATION	EMINENT HISTORIANS	RELEVANCE TODAY
Bright lights, big city	Platt: *The Electric City* Isenstadt: *Cities of Light*	Street as platform, love of responsive glow
Edison vs. Tesla	Jonnes: *Empires of Light* Freeberg: *Age of Edison*	Inventor culture, disruptive technologies
Grand social projects	Hughes: *Networks of Power*	The great transition (beyond fossil fuels)
Monopoly and policy	Lambert: *The Power Brokers* Bakke: *The Grid*	Gilded age 2.0 vs. energy democracy
Fears and superstitions	Marvin: *When Old Technologies Were New* Simon: *Dark Light*	Smartgrid utopia and dystopia
Built environment	Banham: *Architecture of the Well-Tempered Environment* Nye: *Electrifying America*	Responsive buildings, good city life
New Deal in the home	Tobey: *Technology as Freedom* Pence: *The Next Greatest Thing*	Smart home, energy districts
Automation	Winner: *Autonomous Technology* Nye: *America's Assembly Line*	Robotics and sensate spaces
Modernity	Lieberman: *Power Lines* Mumford: *Technics and Civilization*	Cultural expression, neo-technical arts, and language

2.3   Prominent historians and themes in the social history of early electrification.

Long reads on historic lessons for grid awareness reach ever wider audiences, and one, by the anthropologist Gretchen Bakke, became a best seller and a Bill Gates's pick in his top books of the year (2016).[10] Social historian of technology David Nye has devoted a career to works on the distinctly American ethos of energy abundance, material productivity, and the technological sublime. The academic discipline of science, technology, society has frequently studied the process of electrification to illustrate the importance of cultural position and the need for periods of interpretative flexibility.

Throughout such writings, streetcars or assembly lines may not be the foremost image of early electrification, however. More often that is city lights. As so many historians are rightly fond of retelling, electric light made its public debut at the grandest scale in world's fairs, first and foremost the Columbian Exposition in Chicago in 1893.[11] Relative to its smoky, fire-risking predecessors, whether something so ancient as a torch or a fairly recent thing like a kerosene burning lamp, electric light was indeed strange. Although many visitors had encountered individual electric devices, few had witnessed electrification on so grandly dramatic a scale. The nightly illuminations on the grand court of honor were the hot ticket at the exposition. The great risk of fire in a crowd was gone, and in its place was a brighter but cooler phenomenon that could be activated, redirected, colored, or faded in an instant. Unlike the familiar sight of light emanating from a flame, now projected light bathed entire buildings from afar. In a widely recognized irony for so progressive an event, the fair had a neoclassical architecture, the kind best (from centuries of practice) at modulating how light falls across a facade: from above. But much as even a familiar face would be, when lit from below it became strange. This strangeness helped visitors believe they were seeing electricity itself, if by means of its applications in lighting. This proved prescient for how electricity would animate and defamiliarize so much else about everyday life while assuming the form of its host architecture.[12]

Then in 1896, master entrepreneur and electric power tycoon Samuel Insull opened a new Chicago power plant on Harrison Street, the largest in the world at the time, producing power at half the unit fuel cost of its predecessors, and using the very engines and generators that had powered the 1893 world's fair.[13] That same year, Tesla and the Pittsburgh industrialist George Westinghouse opened the world's first hydroelectric plant at Niagara Falls, and from that to Buffalo, the first full-time intercity transmission line. These established the means as well as the model, if not yet the widespread reach, of what would become the grid.[14] Many decades would pass, and strong government intervention would become necessary, before all America (including rural Tennessee) could connect. In the meantime a cultural divide widened, and not without superstitions.

Strangeness

In a manner much like newly sensate media today, where processes not understood are casually called "smart," the novelty of electricity tantalized and mystified. This episode may well be of relevance for today's uncanny wonder at a more interactive, digital responsiveness.

Looking farther back in the history, first it is important to acknowledge the pure science. Physics has few formulations so elegant as James Clerk Maxwell's unified field mathematics of electromagnetism (1862), nor demonstrations so foundational to so much future engineering as Michael Faraday's first prototype of the fields and currents from moving coil technologies (circa 1830). It also worth noting, however, that these breakthroughs encouraged all manner of field theory speculation.

Even so late as 1910, by which time electricity had become an everyday phenomenon at least for wealthy urban citizens, the award-winning British mathematician E. T. Whitaker published *A History of the Theories of Aether and Electricity*. Aether meant quintessence: "It [was] thus erroneous to regard the heavenly bodies as isolated in vacant space; around and between them is an incessant conveyance and transformation of energy. To the vehicle of this activity the name aether has been given."[15] This was not just folk superstition, although electricity met with plenty of that, and still does. Whitaker's treatise also characterized the period's highbrow interest in a nonmechanical metamedium, as anticipated by such philosophers as Henri Bergson or William James, and later interpreted by the physicists Hendrik Antoon Lorentz and Albert Einstein. As explained by Linda Simon, a leading biographer of James, the last quarter of the nineteenth century (in America) provided "a time when electricity was a stronger force in the imagination than it was in reality."[16]

With this came a surge of not only interest but also apprehension. For as many historians have observed, factories were slow to give up steam-powered drives, houses continued to be lit by gas burners, and public uses of electric light continued to be approached as spectacle far longer than the technology alone would have suggested. "Since the light bulb seems like an incontestably good idea, and since we ourselves would be horrified at giving up electrification, I wondered why the nineteenth century public needed so much persuasion to try electricity," Simon observed.[17] At least not in the home; instead, many strange first encounters were in the body, such as electroshock therapy, jewelry that lit up, electric belts touted to enhance sexual vitality, and uncanny

X-rays that peered into living flesh. Interest and apprehension centered on how, without precedent, the body could now be moved about, comforted, overshadowed, or even killed by electric technology.[18]

To almost anyone who first engineered, applied, encountered, and distrusted electricity, a strong sense of wonder came from activation itself. The strangeness of electricity made it a bit too easy to conflate with an imagined life force. Earlier technologies of water and steam had activated objects long before electric power did so, but never so invisibly, instantly, or controllably. Mechanical feedback controlled systems long before electronic sensors and actuators, but never with such precision, programmability, and memory. In place of a dancing flame, animism was now an invisible flow, a magical level of control, and a dangerous momentary spark.

Note some parallels with today. Early electricity's more fantastic aspects illustrate today's widespread credulity about sentient objects, buildings, cities, and grids. Like "smart" today, "electric" described just about any new entity whose workings were autonomous but not obvious, and whose cultural position was progressive, or at least possibly profitable. The practicality was incontestable, perhaps first in particular industrial and commercial works, but soon enough in a proliferation of new products and settings in everyday life too. Like digital apps and augmentations today, this new mediation became a mark of social standing and in many amusing, often-wearable forms, a fashion accessory. Electricity "had a prestige attached to it as a vital force in human affairs," Nye observed in his influential inquiry into the social construction of "what was electricity."[19]

Whether economic force, life force, or simply the driving force of so many new urban amenities, electricity was the avatar of modernity. Like smartphone apps today, its novelties became symbols of new time, new outlooks, and new habits. To Nye, electricity "played a central role in the creation of a twentieth century sensibility."[20] And it was strange. Practical everyday use did not soon relieve that, especially for those who went without. As Hughes later emphasized, industrial design and marketing learned to play up the revolutionary aspect of what was not only technological but also social, cultural, and psychological change.[21] As countless travelers (and many historians) have observed on this spot ever since, the popular attitude was reflected in Harvard president Charles W. Eliot's aphorism inscribed over Union Depot in Washington, DC, completed in 1907: "Electricity: carrier of light and power; devourer of time and space; bearer of human speech over land and sea; greatest servant of man—yet itself unknown."[22]

Usability of the Modern City

Note the relevance of architecture and urbanism. The activated life was an urban life. Like mobile media today, electricity made the city more usable. Consider some of its contexts and social practices. In particular, imagine the varieties of experience at street level. Although civic planners took great interest in the prospects, and policies at all levels shaped the implementations, the everyday reality was chaotic. Although the grid later became a great public work, with little doubt left about what it was and who was to run it, that was not the case earlier on; at this stage it was mostly run by private companies, in private specializations and property developments. Their overall trajectory was toward abundance. Altogether they provided more citizens with more amenities than any previous urbanism had done. But it was never thought to be for everyone. For those who had access, electrical urban amenities made the city more useful, amusing, safe, and convenient. In some areas, to at least some eyes willing to enjoy the chaotic new contrasts, it somehow also made the city more beautiful. On the whole, electrification accelerated the process by which the incontestable prosperity of the turn-of-the-century American city began to relieve the horrific industrial squalor that had accompanied the rise of that wealth. For instance, in one application that did soon provide universal benefits, electric pumps enabled citywide water and sanitation services.[23] (Water pumping remains among the most healthful, laborsaving applications of electricity.)

Most of this usability arose at individual urban sites and in distinct new categories of architectural context: the office tower, the department store, the five-cent theater, the amusement park, the assembly-line factory, and the domestic economies of the electrified bungalow (figure 2.4).

Consider the aforementioned Jazz Age skyscraper in a bit more detail. In one famous instance, President Woodrow Wilson pushed an electrical button to light up the Woolworth Building on its opening in 1913. For what was the tallest building in the world at the time, a typical floor plan featured a lobby for twenty elevators—a prominent encounter with electrical modernity (figure 2.5). Of course the skyscraper took more than electric elevators to become a workable building type. Electric lights evened out the high daylight exposure. Telephones (another kind of wiring) let larger tenants spread over several smaller floors, and let them work in an office tower far from their factory or logistical sites.[24] Electric water pumps provided better fire protection and modern restrooms high above city reservoirs. Fans moved air. Signal beacons beamed from the spire (often about weather, not yet the backlit corporate logos for which tall buildings would eventually seem like signposts). Districts of these office towers, whose density would not have been practical without

Skyscraper
elevators, water pumps, telephones

Assembly line
precision tools, material handling

Amusement park
access, sounds, rides

Bungalow
kitchen, radio, lighting

5-cent theater
movies and air conditioning

Department store
displays and escalators

Transit hub
trains, light, air, and signage

2.4 Architectural types of the electrified city.

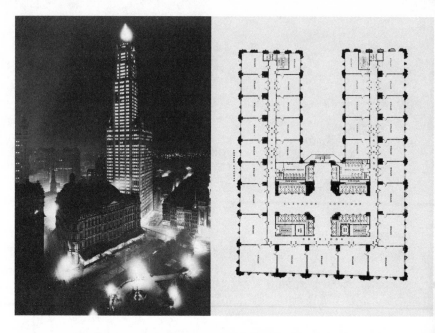

2.5 Electric technology at the center of the Jazz Age skyscraper: a typical floor plan of the Woolworth Building. Cass Gilbert, 1914.

electric subways, created a new density of transactions and exchange, and with it a new wealth increasingly removed from factory and waterfront, increasingly applied toward accessories of modern life.

In the department store, another emblematic building type of the streetcar era, accessories were on offer in glamorous new contexts. Electric light not only allowed large, deep floors, far from exterior windows, but also could be designed to create theatrical appeal in featured displays throughout. Escalators kept customers moving and so added to the sense of browsing. Cash registers began automating the point of sale (still a hot topic in interaction design today). Air-conditioning had some of its first nonindustrial, inhabitable applications in department stores, although not until the close of this early era. Radio sometimes filled the air too, and radio receiver sets appeared in appealingly lit retail floor displays.

In 1924, a young Lewis Mumford, who went on to become a prominent voice of a "neotechnic" era, wrote in his influential early book, *Sticks and Stones*, "A modern building is an establishment devoted to the manufacture of light, the circulation of air, the maintenance of a uniform temperature, and the vertical transportation of its occupants." Electrification had led to a science of air. "[Engineers] hold that the maximum efficiency demands the elimination of windows, the provision of 'treated' air, and the lighting of the building throughout the day by electricity," Mumford marveled. "All this would perhaps seem a little fantastic, were it not for the fact that we have step by step approached the reality."[25]

Despite the glamour of city lights, perhaps the most characteristic social pattern of the era was mass production. As electrification enabled the assembly line, it removed (to distant power plants) the noise, heat, and smoke of furnaces and boilers, formerly such detriments to factory work. Air-conditioning had its first applications in temperature-sensitive industrial processes. Pumps moved large volumes of water on demand. Electrification improved conveyor belts to move material to workers. It also enabled new materials, particularly stainless steel and aluminum. It brought new speed and precision to existing machine tools, such as presses and lathes. Elegance of process soon drove a new discipline as well as cultural aesthetic of machine age art and design.[26]

Insights of process discovery and material economy raised a new discipline of industrial design. As a plenitude of products came off the assembly lines, cultural sensibilities shifted. Industrial engineering, long most evident in problem solving at a monumental scale, now applied to desires, conveniences, and everyday tasks.

In architectural historian Reyner Banham's astute characterization (written in retrospect in 1960), a new machine aesthetic arose for "the age of power from the mains and the reduction of machines to human scale."[27] The "main," as a noun, can mean the largest pipe, the whole utility system, or in a usage worth noting with respect to architecture, the place where the resource enters the building. There, the "reduction" enabled a proliferation of new applications.

As testimony to social customs, one especially popular early electric appliance in the new abundance of products was the clothing iron. This replaced an earlier technology that had to be warmed in a fire, and so had capacity to maintain or regulate its temperature. In the everyday life of early electrification, such apps had the longer name "appliances." By contrast, other appliances had little precedent. A vacuum cleaner was not simply an electric broom. Images of the rotary electric fan suggest how unanticipated some of these things were (figure 2.6). An especially famous, oft-cited photo of a fan powered from a light socket reminds that this was an appropriation, or what today would be called a hack, and not a direct, linear development of an inevitable technology. This comes from before standard plugs and sockets, which in America date from 1917. By contrast, a handsome design by Peter Behrens (1908) illustrates the new aesthetic and the prospect of industrial design.

Incidentally, by contrast, Behrens also designed a great work (and a standard citation ever since Banham) for the making of the mains themselves. The AEG turbine factory (Berlin, 1912) still stands today as a monument to the functional aesthetic of producing electricity. Today when most designs appear outwardly as big blank boxes and power plants mostly stand for carbon-spewing obsolescence,

2.6 Personal comfort app: AEG electric fan, 1908. Peter Behrens / Darmstadt Artists' Colony / Wikimedia Commons.

it takes some perspective to understand what a wonder they were. "Nothing is more beautiful than a great humming power-station . . . synthesized in control-panels bristling with levers and gleaming commutators," the Italian poet Filippo Tomasso Marinetti wrote in the "Manifesto of Futurism" in 1912.[28] Consider the famous painting of a power plant (figure 2.7) by Italian futurist architect Antonio Sant'Elia as an emblem of the strange cultural resonance of electrification.

While perhaps not as sublime as its massive dynamos themselves, the control rooms (figure 2.8) of early electrification still provide a vivid cultural image. There the dispatchers ran the daily network operations. Telecommunications not only played a major role but also frequently shared physical corridors and lines. Physical switching and signaling were

2.7 The Power Plant, by the Italian futurist Antonio Sant'Elia (1914).
Wikimedia Commons.

(a)

(b)

(c)

2.8 A control room from each of three eras:
(a) Commonwealth Edison. Scientific American, 1914;
(b) University of Michigan. Bentley Library, 1969; and
(c) California ISO. Alamy, 2013.

already a familiar part of this; relays, for instance, had been around for decades in railroad technology, and now new telephone switchboards were using them too. The mechanical feedback control mechanisms (for example, "governors") so essential to steam technology helped establish the idea of industrial control systems. By the 1920s, the control room had become a normal feature of most large industrial operations. As factories had electrified themselves, many had developed relay logics to not only activate but also implicitly represent the state of particular processes. The state of these could be sent to indicator lights in the control room and changed by remote switches there. Mechanical drum chart recorders provided data visualization over time. Any numerical analysis was still by hand, as were most of the decisions to turn something (often a mechanical steam valve) on or off.

American architecture critic Montgomery Schuyler (1843–1914), who wrote admiringly of utilitarian structures such as bridges, had little to say about electric power infrastructure, which developed late in his lifetime. Talbot Faulkner Hamlin, a later, lesser-known critic, wrote in 1926 that "industry means power; and power houses have received more architectural study than any other industrial buildings."[29] Yet Hamlin's sensibility was of the beaux arts, the prevailing architectural attitude in the early century, which brought a neoclassical elegance to civic amenities, most recognizably to the great museums and train stations still admired today, but also to waterworks and power plants, as if for reassurance, and of course to indicate benign authoritative control, for sites that the public would see but never enter.

Instead, the cultural energy was in making use. As products served an unprecedented range of tasks, and electrified venues enlivened an unprecedented range of activities, they also slowly made the industrial world more knowable. Embodiment in everyday life brought more cultural change than any infrastructure built from the top. Participation, and not just automation, advanced the new sensibility. There is no aesthetic, nor much shift in social practices, without habitual engagement. As Banham so influentially observed, electrification overturned earlier oppositions to industry. "Under these changed circumstances, that barrier of incomprehension that had stood between thinking men and their mechanized environment all through the nineteenth century, in the mind of Marx as much as that of Morris, began to crumble."[30]

Then it could enter the home. By the 1920s, most American cities had a dominant electric utility company, with an easy process for new customers. In 1929, Chicago's Commonwealth Edison hooked up its millionth meter. The price per kilowatt hour had fallen by a factor of four in twenty years. Billing rates were now based by room and no longer by the count of outlets, letting builders install enough outlets to give new construction a new kind of distinction. In the home, the number of electric lights was a fairly good indicator of status.

Once no longer a source of apprehension, electrification became a source of home economics. Historians also like to study how (much like informational apps today) new electric appliances brought scientific management attitudes into the home, and with them, attitude shifts about such fundamental cultural traits as comfort, time management, gender roles, and social standing. So the familiar conundrums arose of time spent on supposedly laborsaving devices. For instance, instead of spending less time on laundry, people washed their clothes more frequently, after fewer times worn.[31]

Radio, an urban phenomenon because of the need to broadcast within a limited range from a central location, connected the home to the city in a vital new way. When the Radio Act of 1927 stabilized that new broadcast industry, services became more reliable, listening became more fashionable, and the desire for a radio receiver brought electricity into many more homes, and often doubled electricity use where that had already been installed. Similarly in the 1930s, the National Recovery Administration promoted the widespread adoption of the fridge, which again doubled domestic electric use.[32] It also promoted homeownership, particularly of "electrically modern dwellings," as a new basis of financial security amid the Depression.[33]

When, in the era to follow, domestic electrical conveniences reached many more people, it eventually became easy to assume how that was always so. In an especially rich study of domestic electricity in the 1920s, Ronald Tobey effectively challenged the widespread belief that this consumerism rapidly reached almost everyone, however, or that the free market ever would do so. In 1920s' Riverside, California, the site of Tobey's case study, "the homes of the elite shown brightly from the hill."[34] Instead, it took government policies to boost access and eventually achieve almost universal reach (in the United States). These belong to the emerging era of consolidated universal service.

Travel electrified too. Besides balancing the load of the local grids, electric trains let cities begin to rid themselves of the soot and smoke of coal-fired steam locomotives. This also altered the urban form. Clean power let trains travel underground. Transfer stations were more safely and comfortably lit. Signage became readable at night, and from afar. Regional travel revealed the growing contrast between the electrified cities and their still-dark hinterlands. Streetcar suburbs soon offered new housing types, such as the bungalow, and new domestic comforts based on built-in electrification. But for many existing neighborhoods and almost all rural places, those appealing prospects beckoned only from afar. Seen from a distance, bright lights set the modern city apart from its candlelit hinterlands.[35]

Consolidation

Seen in retrospect, it was right around a century ago (the late 1910s) that most of the pieces were in place to create a larger, more universal power grid. At a local urban scale, the era of early electrification had worked out all the components that the later era then consolidated at a national scale. Much as the first quarter of the twenty-first century has brought a surge of innovation in urban informatics, thus consuming ever more data feeds, so the first quarter of the twentieth was for urban electrification, thereby consuming ever more electric power. If not innovation, the era could just as well be named for expansion. In counterpoint to today's turn toward more micro power schemes, this phenomenon of scaling up deserves a bit more narrative here.

The topic of scale became essential fare in the social history of technology, and nowhere more so than for the grid. The bigger the turbines, the greater their efficiency. The more that the industrial, commercial, and increasingly residential public could be induced to consume, the lower the price per unit consumed. The more apps (then called appliances) conceived and sold, the less contrived any one of them seemed. It was not unusual, across so many decades of such steep growth, and even so late as the millennium, for an average American household to use as much electricity in a single month as its grandparents, five or six decades earlier, had used in an entire year.[36]

The biographies are many on how Insull worked natural monopolies to build the public utility model that still prevails today. It would be difficult not to make at least some mention here, since the foundation he built remains, a century later, the one that most investor-owned public utility leaders operate on. The time span of the early electrification era, which worked out at an urban scale all the pieces that the later era then consolidated at a national scale, is just exactly the career of Insull, who as a young immigrant became secretary to Edison in 1881, and as a power broker lost the fortunes of thousands of investors in the crash of 1929, after which he was indicted and died out of favor. In most tellings, the story of Insull is one of natural monopolies. "If you will bring your price down to where you can compel the manufacturer to shut down his private plant because he will save money doing so," Insull wrote in 1915; "if you can compel the street railway to shut down its generating plant; if you can compel the city waterworks . . ."[37] A general-purpose network could balance its many different loads throughout the day: streetcars at dawn, industries around noon, businesses in the afternoon, residences in the evening, and streetlights at night.[38] A large-scale infrastructural investment could justify public policies to protect it.

To obtain the rights on natural monopoly, and avoid the costs of building redundant territorial networks or waging price wars, the companies accepted strict regulation, under which they agreed to provide universal, nondiscriminating service at a reasonable price. These regulations were an advantage, not a burden, because of the enormous capital costs of scaling up. As energy law historian Jeremiah Lambert has explained, "Insull believed that state regulation would reduce the risk of investing in electric utilities, make utility stocks and bonds more attractive, increase the availability of capital, and lower its price. He was proved right on all counts."[39] Insull saw the potential economies of scale sooner and better than anyone else, and pioneered the technology, economics, regulation, and consolidation that made so many local services into one grid, which could scale up endlessly. According to both the physics and economics of the ever-improving turbines, bigger was always better: greater volume always meant lower unit price. As Lambert noted, "Insull created the organizational and technical foundations of the nation's electric utility industry, still largely in place despite decades of intervening change."[40] This was not only entrepreneurial but ultimately cultural. "[Insull was] a devoted believer in the gospel of consumption that set in motion 'a self perpetuating cycle of of rising use and declining rates the eventually enveloped an urban society in a ubiquitous world of energy.'"[41] Or as the Institute for Energy Research has memorably put it, "Samuel Insull did for electricity what Henry Ford did for the automobile—he turned a luxury product into an affordable part of everyday life for millions of Americans."[42]

Central and Universal

"We must work away from the point of view where we use it sparingly." In the 1933 federal plan for rural electrification, the influential environmental utilitarian Gifford Pinchot expressed this aim. The consumption of electricity had been doubling every decade, but was still nowhere near universal. The plan was to accelerate the growth to where everyone could benefit and electricity would resemble that other most fundamental flowing utility: water. "No intelligent person spares the use of water as a means to comfort and cleanliness—even though it costs enough to warrant its being metered. Water as a factor in our lives is free—not so free as air—but so free as to permit us to partake fully of its benefits."[43] So too for electricity.

Through the decades to follow, growth continued to improve the economies of scale. In an upward spiral, more demand supported larger power plants that delivered a lower unit price that induced more demand. As electric power service grew beyond locally specialized use and became a common pool resource, the need for public administration took form. Under increasing government oversight, redundancies were minimized, interregional exchanges were standardized, interconnections helped balancing, loads were better forecasted, and the long-term futures necessary for such enormous investments were stabilized. This was the middle era of "consolidation."[44]

As historians from almost any perspective explain, this set of practices defined an era. The questions are of policy and law, more so than of technology or product design. The results are still ubiquitous, whether in the physical landscape, the managerial attitudes, or the legal jurisdictions. So let that middle era of consolidation receive just this shorter mention in this project. The eras before and after invite more renewed interpretation, at least from the perspective of the built environment, especially since less prevalent in many histories of electrification.

"Of the great construction projects of the last century, none has been more impressive in its technical, economic, and scientific aspects, none has been more influential in its social effects, and none has engaged more thoroughly our constructive instincts and capabilities than the electric power system."[45] So Hughes opened *Networks of Power*. Or as Bakke began her more recent guide, "The grid isn't just some contraption wired together out of various bits of this century and bits of the last. . . . It is also a massive cultural system."[46]

No longer subject to the financial raiders and speculators who had crashed in 1929, now national in scope, standardized across a noncompeting set of regional operators, starting from the 1930s onward, it became much more unified. This too accelerated its growth. Here is where it became "the" grid.

Although federal intervention in any industry remains an anathema to free market politicians (and robber barons), the federal regulation of regional power operators remains the fundamental organization of the US electric grid. Economies of scale developed in Insull's private enterprises performed as well as or better in the New Deal's gigantic public works. Investor-owned public utilities had become predominant much earlier, and remain so (precariously) today, but in this middle era of utmost centralization, they became pieces of a much larger, increasingly uniform, universal, and (to all but its operators) invisible system. While the heroic projects of the 1930s and especially the huge hydropower dams still make good reading, and their photos remain some of the most iconic images of the energy industry, nevertheless by the close of this era, the public utilities were anything but glamorous. They had succeeded at disappearing from cultural view.

It is worth repeating that unlike today, the authorities fundamentally agreed on how it all worked, and who was to work it. These agreements and assumptions held firmly for half a century. That interval was long enough, in an otherwise astonishingly changing world, to establish the electric power business as a haven for the most conforming, risk-averse engineers, policy makers, and investors that society could muster. That too made the field into something everyone else would prefer to ignore.

From the perspective of the built environment, the provision of not only power but also its uses and effects become increasingly standard. The more electricity that was applied to inhabited space, the more it all seemed alike. An overly narrow consensus developed around building and dwelling too. Today such twentieth-century consensus has become an outdated, restrictive, legacy to escape. All too familiar, it too needs rather less interpretation here. Thus for nearly a century, up until today's flood of change, everybody knew what electric power systems were: rapidly growing, centrally generated, regionally administered, investor owned, and publicly owned, with minimal redundancy and huge economies of scale, under which it always made sense to make more, and to do so in as centralized, standardized a way as possible. It is no surprise how today's many corporate notions of the smart city tend to perpetuate that approach.

So those are three good reasons why relatively few scholars have chosen to investigate grid awareness from a perspective of architecture and urbanism. And yet now change has come, and choices lie ahead.

Diversification

To many of its advocates today, the grid seems like the last, biggest, most foundational technology to be remade by the internet era. By now it seems too brittle, too antiquated, and too poorly prepared for any longer emergencies ahead. To other, more technoutopian futurists, the very idea, and especially the gigantic scale, of centrally generating and transmitting electricity can seem antiquated. Some utopians appear ready to imagine newer wonders of activation, running on superabundant nanoenergy materials, wirelessly and invisibly built into any object as its own source of power. To the pragmatists of the present, in an update of the usual metonym of keeping the lights on, the highest priority is to keep the internet on. Apple and Google have been experimenting with local, renewable power for their server farms. Indeed, many critical organizations like hospitals, military bases, and research labs, plus a few municipal experiments in district-scale design for living, now turn toward the micro scale.

Microgrids become the latest stage in the social history of electrification. The next greatest thing to having electricity in your house is to have electricity when most of the city does not.

So for wisdom in important choices ahead, it helps to take a long view on electrification. After all, electricity itself is still fairly recent in any longer prospect of humanity, architecture, and design for living. That makes its present transformation by information networks all the more due for more flexible cultural interpretation. It could be a mistake to let entrenched organizations treat this simply as grid "modernization."

Although a vast and delightful historic literature exists, nevertheless the academic literature remains fairly thin on these more contemporary interests. Much of the recent literature is from corporate infrastructure builders instead, too much design literature tends toward technoutopia, and meanwhile the popular trade press tends toward prepping for apocalypse.

So with a much more modest agenda, let this project seek a small contribution to the science, technology, and society literature of the microgrid boom through the lens of architecture and urbanism. Let this address design for everyday living, as if the relationship of the built world and local energy becomes worth considering again, and as if some of the engineers, economists, and policy makers who more usually debate reelectrification might take some occasional interest in the built world themselves. As for designers, an aesthetic might conceivably arise: new components, new configurations, and even new urban forms, even if none of them quite so prominent as Jazz Age New York skyscrapers.

3 SMART GREEN BLUES

Here in electricity's latest era, inevitable change has come. As briefly noted in opening, several simultaneous trends now destabilize an electric power industry that had worked so long to avoid just that. Thus a new grid awareness must expand. These changing attitudes make good fare for science, technology, and society studies. That in turn deserves a fresh approach from the disciplines of architecture and urbanism. Those bring insights not only into local scale and identity but also cultural speculation. Often under the name of *the smart city*, today's cultural imagination welcomes local energy, sensate buildings, interactive participation, and resilient urban archipelagos. Each of these themes expands grid awareness. Thus across several respective chapters to follow, that is the arc to the argument ahead.

How does a work in science, technology, and society take up such a narrative? Since that field often emphasizes history, this work has just provided a brief such overview. But then since that field has such enduring interest in electrification, surely it needs many new works on more the recent conditions that have been summarized here as, "decentralization, decarbonization, Internet of Things, and external threats." Even more so than it investigates history, the field of science, technology, and society emphasizes the cultural dynamics of technological change. Here, then, consider the rise of an increasingly *hybrid forum*. In particular, consider an increasingly cultural perspective on what is otherwise assumed to be a mainly technical challenge. To take up this narrative, stay with one simple proposition: *a new grid awareness* leads away from the kinds of utopian totalities often imagined as *the smart city*, toward more convivial, resilient adaptation at local scale.

Thus to extend the theme of electrification's eras, consider the ongoing era of decentralization, less as historic storyline and more as contemporary witness to sociopolitical dynamics. To do so demands some taste of the issues' complexity but benefits from staying at an altitude high enough not to get mired in the technical details. It takes a bit more exposition of the general conditions at hand, and does so with a mind that many more disciplines now seek their respective takes on a situation that few should ignore.

To jump back in to that general scene, zoom out to planetary scale, where strategists see a "Great Transition" beyond fossil fuels now accelerating. To Lester Brown, the eminent founder of the Worldwatch and Earth Policy Institutes who coined that phrase, this change has become not just inevitable but indeed definitive: "The shift from coal, oil, and natural gas to solar and wind energy will be the defining event of our era," Brown declares. This is an optimistic trend toward "a degree of personal energy independence not known for generations."[1]

The prospect has quickly become normative. Hence the era of decentralization: today anyone might own, operate, sometimes trade, and perhaps culturally identify with *clean green local micro* electricity. Today in many regions it is already economical (or "grid competitive") to do so. As Brown explains with a book full of recent data, this is no longer just wishful thinking. For instance, "even oil-exporting Saudi Arabia [has enough solar generation under planning and construction] to generate ⅔ of its electricity from the sun."[2]

With locality as the theme, consider the players. Amid such global change, the scene of cultural dynamics becomes the neighborhood, the town, or the region. To civic activists, clean energy becomes a topic of local ownership and "community choice" (figure 3.1). To city planners, resilience has ever more value and is almost always local. To financiers, the network advantages are usually in the aggregate as well as in the dynamics of localized virtual markets. To the major infrastructure builders, who must be recognized here as the main voices of technological change, innovation increasingly occurs out at the grid edge.

So to resume the opening argument on edges, change is now catalyzed by pervasive digital design. In contrast to the massive dynamos and big transmission lines so prominent in twentieth-century thinking, today the millions of smaller devices interlinking across the grid have become its main locus of change. To the designers of such devices, the Internet of Things raises expectations that many other objects and environments not primarily thought of as digital devices can sense, process, communicate, calibrate, and learn. To some strategists the electric grid is itself one of those remade things; some call it the largest thing on the Internet of Things.[3] Using tech-release jargon, many strategists now call it all "Energy 2.0."

What localities see as a bright green future, though, often appears much more grim at the center. Noted briefly in the opening, this likewise deserves some unpacking. As seen in longer historic perspective, century-old economies of scale have reversed. Although this has less direct implications in architecture and urbanism, it is inseparable from edge developments, as from any wider grid awareness. Large-scale

3.1 Community choice: a banner for local electricity, in a style to recall early rural electric cooperatives, from Sonoma Clean Power, a contemporary pioneer at Community Choice Aggregation. Courtesy of Sonoma Clean Power.

centralization still has real advantages for balancing variable supply and loads, maintaining overall power quality, or interconnecting complementary resources and regions. Yet growth has slowed, and resources have aged. To conventional grid operators, this means trouble. For example, as Brown points out, "The two largest German utilities, E.ON and RWE, each saw their market value drop by more than half between 2009 and 2013."[4] Here in North America too, operators face increasing costs to maintain decaying infrastructure, but decreasing surplus from which to fund reforms. Their economies of scale erode as users defect to independent local generation. With fewer customers left to bear shared infrastructural costs, the utility operators must increase their rates, which induces still more defectors. Built on a century of growth spirals, utility companies now express fear of "death spirals." Under these circumstances, only the most entrenched operators speak of "grid modernization"—that is, updates without reform to the existing worldview as well. Grid modernization also frequently proceeds as if it is culturally value neutral. Something more dramatic must occur. Given the growing value of resilience, it is mainly a matter of how.

Open Cultural Imagination

Any cultural understanding of technological systems goes through cycles of open exploration and conceptual closure. Electrification's eras have done just that. The disciplinary lens of science, technology, and society sharpens such understandings. Where others might simply declare a "paradigm shift" as if a singular event, social historians of technology investigate how hybrid forums open and close debates over time.

Consider some such language of cultural dynamics. Within such cycles, early periods of innovation thrive on *interpretative flexibility*.[5] Where such flexibility brings success, however, a stage of consensus then arrives. As expressed in classic sociological jargon, "controversies are terminated" by disciplines' "closure mechanisms."[6] Sometimes laws and policies set the course. Sometimes the closure occurs through membership in an elite. Under agreed-on criteria and reproducible formulas, expert analyses dictate this and not that. Still more becomes out of the question. Soon these closure mechanisms translate into a "wider social milieu."[7] The criteria used by the managing elite become normal, acceptable criteria to everyone else, who henceforth assume them and eventually lose awareness of having done so. High modernist central electrification, in its consolidation, was such a period of conceptual closure. By contrast, early electrification was a chaotic period of interpretative flexibility. From today's perspective amid a new stage of change, that earlier era appears the more interesting one—again worth a long look by many more fields—and the middle era still appears as the more familiar legacy to escape.

Forty Years in the Making

So to extend the historic background for a new grid awareness, briefly note how a series of events (figure 3.2) from the present lifetime have led toward such remarkable change in so fundamental a resource. Although hardly so dramatic as the earlier transition from innovation to consolidation (around the stock market crash of 1929 and the controversial New Deal programs of the 1930s), nevertheless around forty years ago another era began. Starting around 1980, and still gathering momentum today, the destabilization of the electric power industry has become difficult to ignore, predict, or by now, leave entirely to others.

As recognized best in another landmark of science, technology, and society, written by the historian Richard Hirsh in the 1990s, the prospect of a new era started as the breakup of a "utility consensus." In terms of a single most significant breaking point, Hirsh identified the Public Utility Regulatory Policies Act of 1978 as the beginning of the end of "an agreement based on the widespread belief that the electric power business constituted a natural monopoly." Across that previous half century since 1929, an elite managerial class had gained dominance, built out the networks according to its formulations, and operated them for maximum stability. But now it encountered problems more heterogeneous than its homogeneous composition could address. These Hirsh characterized as three main stresses: "technological stasis, the energy crisis, and the environmental movement."[8] First stasis: when the scaling up of thermal power turbines at last met with the physical limits of thermodynamic efficiency, their decades-long growth spiral soon stalled. Then in the energy crisis, fuel costs rose, and the previously unconsidered prospect of conservation arose. Meanwhile with environmentalism, the perpetual growth of induced

1970s	*Energy crisis*
	Energy Act
1980s	*Market deregulation*
	Nuclear boondoggles
	Intelligent electronic devices
1990s	*Building management systems*
	Internet trading
	Grid reform literature
2000s	*California brownouts*
	Wireless sensors
	Enron bankruptcy
	Renewable portfolio standards
	Planetary change awareness
	Mobile apps
	Green building standards
2010s	*Culture of perpetual interactivity*
	Net metering and feed-in tariffs
	Utility consumer apps
	Carbon neutral districts
	Community choice programs
	Quantified self movement
	Electric vehicle futurism
	Solar generation cost competitive
	Local energy storage speculations
	Major outages from superstorms

3.2 A time line of the era.

consumption came into question, as did the nuclear reactors that were the latest way of powering it.

After the close call with disaster at Three Mile Island in 1979, and then in the unforgettable meltdown at Chernobyl in 1986, nuclear power quickly lost its role as the consensus technological future. Just as influential, if hardly so dramatic, in 1982 America's largest nuclear construction project, on the Columbia River at Hanford, suffered an unprecedented fiscal default. This too marked the end of a much longer era. Significant for its ties to the Bonneville Power Administration, the showpiece of the New Deal half a century earlier (whose nearby Grand Coulee Dam at six gigawatts was still America's largest hydropower site), the boondoggles of the Washington Public Power Supply System became known by pronouncing the acronym WPPSS as "Whoops!"[9]

Large government projects were on the way out anyway. The Reagan-Thatcher turn of the 1980s brought a dramatic cultural shift toward neoliberalism. From then on deregulation, liquidation, and bottom-line economic reductivism came to almost every field. Fiscal priorities henceforth eclipsed all other priorities, even for highly regulated public utilities.[10] On all these fronts, managerial conformity soon fell out of favor. Instead, new creeds of resource economy, market deregulation, and privatization, catalyzed, for instance, by MIT economist Paul Joskow in *Markets for Power* (1983), now broke open the field's former closure mechanisms.[11]

Then as with so many other new financial instruments that so characterized the 1980s, the abstract wholesale trading of electric power accelerated.[12] When in 1992 the United States passed an energy policy act that overturned legal monopolies that had been in place since the 1930s, and in doing so gave any independent producer the right to transmission at a fair price on any utility's lines, electricity's new era began to come into its own. This opened new market plays on the regional differences so important to grid balancing operations.

The most infamous of these was Enron.[13] Since California remains so influential in the cultural imagination, and since little has done more then Enron to undermine any future trust of smartgrid, this is still worth reciting today. At its peak, just fifteen years into its existence, Enron had a market capitalization comparable to Exxon and was making tens of millions dollars daily in California alone. Progressive, energy-aware California had created daily spot delivery markets, which Enron abusively gamed. By manipulating newly abstract entities of power transmission and delivery rights, Enron worsened volatility and congestion, deliberately produced scarcity, and then made soaring profits off that. By the turn of the millennium, California, the undisputed leader of the internet age, suffered rolling blackouts and as much as an eightfold increase in wholesale electricity prices. Some Californians may never trust internet energy companies again.

By the early 2000s, another cost of reductive emphasis on price alone had become evident: underinvestment in aging infrastructure. When the great Northeast blackout of August 13, 2003, the largest in North American history, left over fifty million people without power, bad fault detection took the blame. As understood in a larger cultural context, this event showed how much technological dependency had deepened while infrastructure had frayed. As historian David Nye went on to examine in an especially worthwhile read on blackouts, "It is not just the electrical system that breaks down; the social construction of reality breaks down too." In August 2003, millions of people rediscovered how infrastructure becomes much more visible on breakdown. "In 2003, New Yorkers were dismayed to find that among the thousands of devices that did not work in a blackout were mobile phones, gasoline pumps, escalators, automated teller machines, and pumps in the water system."[14]

Nevertheless the love of technology still deepens. In what is surely the greatest step into unfamiliar futures that has occurred within the last forty years, smartphones have quickly altered cultural sensibilities. By 2007, the year that Apple introduced the iPhone, new thinking arose about personal mobile apps as a basis for a new urban computing, or as was often then known as "locative media." At the same time, wirelessly embedded sensors and processors were diversifying the range of things being called smart. Soon enough, the smart city became a fashionable speculation. Buildings, networks, and of course grids could soon be imagined smart as well.

In smart green building, for example, the design integration of sensor and actuator systems now made more architectural elements more responsive. The old engineering wish to make the built environment mostly invisible to its inhabitants now seemed a thing of the past. The overstandardized, glassed-in buildings of electricity's previous, profligate era no longer seemed healthy nor enjoyable nor sustainable. The time had come to replace them with something more sensate, sensory, participatory, and noticeable.

Power itself likewise gets cleaner and more participatory. The 2010s in particular have brought wave after wave of increasingly affordable wind and solar power generation. Thus across the previous forty years, and especially accelerating in these last ten years, many aims for a softer, greener, more locally self-reliant energy have gone from naive countercultural fringe to mainstream strategic planning, and even (in some contexts) into everyday practical implementation. Solar power has become the most visible of these, and is often the trigger for related change. Many more policy-intensive principles all have earned vigorous development and debate: carbon pricing, life cycle costing, energy intensity, negawatts, and a more locally institutional approach to resilience.

To take a single recent year as a sample, consider 2016, the year that prompted this project. It was the first year in which the majority of America's new energy generation coming online was from renewables. The overall share of power coming from renewables was still under 10 percent, however. Nonetheless for a variety of reasons, by 2016 it became clear that the nationwide demand for electric power was no longer increasing and had gone into decline in some areas.[15] It was the year in which Elon Musk's provocative technofuturism reached mainstream audiences. It was a year of public debates and ballot initiatives (some of which did go retrograde) over the two-way net metering of rooftop solar power. It was a year of convulsions in the solar energy business: after several years of double-digit growth, there came a residential slowdown as plummeting prices created a glut, and some bankruptcies followed. Solar instead took off at utility scale and in corporate off-site plants. Community choice aggregation rose. Journalists recognized a microgrid boom. Then, as if nothing else would still defend their interests, on 11/9 (2016), fossil fuel billionaires elected a reactionary government. Yet not even that has stopped the Great Transition.

Alongside Brown's oft-cited reporting, several well-respected guides offer clear views of this more recent trajectory. Soft energy pioneer Amory Lovins, who has remained prolific throughout the era, provided his most effective overview in *Reinventing Fire*.[16] Mason Wilrich, a former director of California Independent System Operator (which oversees the state's wide-area power networks), has supplied an insider's framework in *Modernizing America's Electricity Infrastructure*.[17] Energy strategist Peter Fox-Penner wrote the single most readable overview in *Smart Power* (2008), which is still in worldwide use, and from which it has since become obligatory to cite Fox-Penner's famous remark about how smart power seems "similar to rebuilding our entire airline fleet, along with our runways and air traffic control system, while the planes are up in the air, filled with passengers."[18] In a 2015 book, environmental scientists Jennie Stephens, Elizabeth Wilson, and Tarla Rai Peterson provided a deep dive into the political complexities, based on hundreds of interviews, in *Smart Grid (R)Evolution: Electric Power Struggles.*[19] As already noted, Gretchen Bakke, who wrote the most widely read recent work (2016) on new grid awareness, is neither an engineer nor a policy analyst but an anthropologist. In conversation, Bakke notes surprise at how many readers from within the power industry itself have discovered her work. From that she takes encouragement how grid awareness need not seem ideological so much as practical: it is good preparation for times of rapid technological change.[20]

Who Gets to Play?

Today many more players want to own a piece of this bright green energy future. Many more join the *hybrid forum*. Leave it to the major engineering houses to build the components, the fast-moving world of market research consultancies to make any fiscal sense of such volatility, and state commissions to haggle new rules, but as forum after forum has shown, do not leave it all to investor-owned public utility companies.

The engineers and policy analysts who guide the electric power industry do of course have the most experience, most accurate simulations, and advantage of long-established organizations. They have the best performance models, generation portfolios, long-distance transmission, wide-area balancing mechanisms, and interregional regional trading—

all that gets called "the" grid. On the ground, the public utility companies still have the best (and often only) operational control centers, distribution nodes, maintenance fleets, and line crews—all that, in a word, often gets called "hardhats." That is a respectful epithet for people working with a dangerous resource. Yet if it ever was so, the grid's future is not just a hardhat challenge.

Alas today, where trust generally erodes in just about any form of centralized power, whether electrical, political, or institutional, and where underinvestment in anything public can soon catch up with everyone, the public utilities no longer hold such esteem as they did half a century ago.[21] Today the public utilities are no longer the only ones providing power; no longer necessarily remain the most innovative builders of energy infrastructure, especially its new layers of information infrastructure; and no longer hold the only legitimate positions in energy policy debate. Their reputation particularly suffers wherever they resist or seek to own all wind and solar generation.[22] By now their monopolies no longer seem so natural, and their great economies of scale have reversed. Yet they strive to perpetuate them.

In their richly ethnographic study, Stephens, Wilson, and Peterson have surveyed the new forum: incumbents versus new actors; perceived costs versus unmeasured benefits; who pays, who plays, and who writes the rules; timescales and spatial scales; different actors' perceptions on decentralization; consumer empowerment versus disempowerment; security or vulnerability; and reduce or perpetuate fossil fuels. They remind all concerned that "to appreciate the critical importance of the social dimensions of smart grid, it is useful to consider the interconnectedness between smart grid and other technology innovations, particularly in communications."[23]

Or as cybersecurity historian Rebecca Slayton has observed, "Thus, the information infrastructure that controls the grid has tremendous potential advantage to some groups at the expense of others—in short, it has tremendous potential for politics." Note how in the years since the most influential academic histories (by, for example, Hirsh or Hughes) were written, information technologies have transformed not only the grid but also much more dramatically, the politics. Since so much politics is local, this invites new ethnographies of space and place. As Slayton suggests, "Historians have yet to explore how the grid's information infrastructure has shaped, and been shaped by, a changing political and regulatory regime."[24]

Well-meaning counterparts in the debate do not lack expertise but just agreement on what most matters. Major private infrastructure builders flood the net with proposed technological (alas, often solely technological) solutions. Local, ad hoc organizations counter with civic and social justice concerns. Major environmentalist networks expose retrenched opposition. Economists debate how to value resilience and not just price efficiency. Urban planners advocate governance, not only since so few cities control their own power resources, but also because neighborhood- and district-scale developments are such an important locus of cultural identity. Meanwhile, as alas is too often the story of just about anything lately, distant digital overlords seek to monetize it all on proprietary platforms.

In the spirit of science, technology, and society, note that whenever separate, well-established fields cannot each resolve a challenge alone, some triangulation from outside their conventional methods can help (figure 3.3). For those respective methods each tend to establish a bias: language, habits, and instrumentation favor one way of designing, operating, or diagnosing a system, usually for the best performance under well-known variables. One name for this

	Information	Energy	Built environment
Major owners	network aggregators	public utilities	individual organizations
Engineers	tech ventures	systems integrators	structural and environmental
Political debate	privacy	regulations	location
Cultural context	social network	universal service	neighborhood and address
Social practice	followers and links	community choice	places of assembly
Rents	data harvesting	ratepayers	tenants
Technological succession	obsolescence	portfolio balance	retrofit
User interface	engaged streaming	set and forget	passive comfort
Experience	activities	power	surroundings

3.3 Triangulating among infrastructures of information, energy, and the built environment, in a qualitative comparison of technological frames that have too often remained separate.

concept is "technological frames." Today, "Silicon Valley" has become a metonym for one such frame.

Silicon Valley Does Smartgrid

The information technology industry thinks it can own this one too. So long as "smart" is the main word for credulous wonder, *smartgrid* remains one of the biggest technofutures around.

For a benchmark on what is usually meant by the term, take an official search-engine-optimized definition from the Electric Power Research Institute: "A Smart Grid is one that incorporates information and communications technology into every aspect of electricity generation, delivery and consumption in order to minimize environmental impact, enhance markets, improve reliability and service, and reduce costs and improve efficiency."[25]

By the turn of the millennium, when the internet had reached into everyday life and the dot-com boom had legitimized technoutopianism, cleantech had become a panacea, and Silicon Valley its most dramatic voice. There is no widespread evidence of a single, authoritative coinage of "smartgrid"; instead, when mobile phones and embedded systems spread computing beyond the desktop into all walks of life, this made it more usual to apply the word "smart" to many more things. For the grid, this soon enough became normative.

In 2007, the US Energy Independence and Security Act characterized a smartgrid. In that same year, legendary Silicon Valley venture capitalist John Doerr declared that clean energy technologies would be "bigger than the Internet."[26] In 2009, the internet infrastructure builder Cisco explained, "The concept of a smart grid emerges from the integration of the power systems view of the electricity grid with its corresponding information systems view." Already that vision involved several terms noted here, such as micro-transactions, end-to-end, and destabilization, for instance. "Smart grid would disrupt the way utilities do business. It would influence the way consumers consume energy and the way they interact with a utility. . . . Cisco's vision is to help create a resilient, self-healing, highly secure, and inclusive grid environment that optimally combines all the disparate sources of information in the utility environment."[27]

A decade in, Cisco now refers to "Digital Utilities."[28] Its smartgrid platform now emphasizes cybersecurity first, besides responsiveness, reliability, and cost/energy savings. Its platform supports identity, communications, analytics, and adaptive learning among an ever-increasing number of devices and networks. Today, where data aggregation platforms have become the largest capital entities of the times, it makes sense that energy would be managed as a platform.

By the early 2010s, Silicon Valley had also embraced the Internet of Things. Under this larger trend, the smart city thus became an informational cloud comprised of countless devices in an "industrial internet of things" (figure 3.4). As the thinking moved toward devices and clouds, and especially toward operations between embedded devices and handheld ones, its physical scale in the city seemed less of a concern. In other words, the city became an extension of internet thinking, more than the Internet of Things came to existing patterns in district energy or urbanist thinking.

To extend the snapshot of the sample year 2016, consider some stars amid the volatility of Silicon Valley doing smartgrid at the time. For example, Silver Spring Networks had become a favorite for the advantage of wireless mesh networks in the field. Technologically, ad hoc and peer-to-peer communications were improving the processes of monitoring, interoperability, and responsiveness, which in

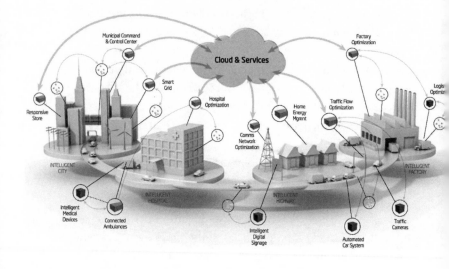

3.4 An illustration of the smart city as a cloud linking the industrial Internet of Things, 2011. Courtesy of Intel.

turn were reducing waste and upkeep. Culturally, this began to give organizations and individuals more control, and hence more awareness. By contrast, Tendril Networks, another star, was emphasizing a more centralized analysis of very large numbers of individual energy users, even as they became aware of monitoring and tuning their domestic usage (figure 3.5). When applied to late afternoon air-conditioning, which is often a contentious topic in grid balancing, Tendril was claiming peak load reductions of up to half in some areas, and overall energy use reductions of up to one-fifth. Networked Energy Services, another prospering company, was instead focused on cybersecurity. After all, as the number of owners, devices, and communications was increasing so rapidly out

3.5 New grid awareness through casual engagement, from analytics leader Tendril Networks. Courtesy of Tendril.

on the edges of an increasingly distributed grid, so was the hacker exposure. Those are just three, but as of mid-2016, Greentech Media, a prominent news and networking organization, counted one hundred companies in its "Grid Edge Council."[29]

By the late 2010s, at this writing, the economic power of information networks was also being realized under the newest major economic category as data platforms.[30] Although the electric power industry is too complex, critical, regulated, and full of dangerous work to evolve so rapidly as the comparatively simpler businesses of Uber rides or Airbnb lodgings have done, nevertheless there are many pieces in place for platforms. Not only may generation occur almost anywhere, so can collective local markets (sometimes known as "community choice aggregation"), aggregate local generation (now known as "virtual power plants"), and with great potential for cultural value change, local ownership and operation by nonspecialists. When companies that would perpetuate the old model run up against companies dedicated to disruption, often both fail, and somebody else materializes in their place.

As today's volatile churn of investors, companies, and services surely indicates, countless new players arise, a few endure, and some of the old standards disappear. For instance, the twentieth-century giant Westinghouse, whose namesake founder was there alongside Nikola Tesla at Niagara in 1896, went bankrupt in 2017.

So in one last historic anecdote here, take a moment to understand how the very notion of smart has had many previous stages in electrification. Whereas historians do not appear to have identified a single, authoritative coinage of the expression "smartgrid," they do agree about intelligent devices in earlier electrical engineering. It is not so much that the grid itself was being called smart, which is a relatively recent usage, as that some of the first electronic things that were being called smart were elements of the power grid.[31] In a work of science, technology, and society on rethinking smart cities, this is worth retelling with at least some technical detail. Generally since the 1920s (and in some instances, earlier), central control rooms had been sending and receiving pulse-coded data to and from remote substation panels to check or activate them.[32] Fast-forward to the 1940s, and the field had become a leader in industrial feedback control, and this inspired formulations of what pioneering computer scientists came to call *cybernetics*. By the 1960s, a standard approach for wide-area industrial process control had emerged. With the benefit of solid-state transistor circuitry and early digital computers, a network could scan and supervise the state of a great number of devices at a great distance—electrical substations among them. With a circuit card present in its panel, each device could monitor its on/off state, and in some cases its volume, and send these over wires to a central computer; hence they were called remote terminal units. A standard framework emerged for supervisory control and data acquisition, and this remains fundamental to grid operations today. Thus in 1964, the *New York Times* reported to its general readership on "the importance of the computer in speeding up the production and distribution of electricity."[33]

Thus there was smart before there was an internet. Then in the 1980s, microprocessors replaced circuit boards, and as these began to integrate scanning, calculating, communicating, and process control in a single entity, the Institute of Electrical and Electronics Engineers coined the expression "intelligent electronic device." In one of the few academic studies of such origins of smartgrid thinking, Slayton has found that "in 1989, Kurt Yeager, the head of the Electric Power Research Institute's Generation and Storage Division, promised: 'The microprocessor revolution will join with and spur all other technological innovations in the next few decades. It is both building block and mortar in creating the Second Electrical Century.'"[34]

So today when Silicon Valley does smartgrid, or mayors tout the smart city, keep such words in mind. In more everyday usage, back from when it was only applied to living beings, the word "smart" still connotes sentient presence. Although embedded software intelligence does make things (or sites or systems or networks) more usable, more interconnected, and more responsive, it also takes open philosophical debates to unpack. Here is the greater ontological debate of the times. Having agency does not necessarily make anything "think," much less reflect, much less reflect on ethics. Not everything that senses is sentient. A simple mechanical float valve demonstrates as much. So do the automatic sensors, switches, and signaling devices in a substation breaker panel. Yet "smart" has stuck. It is at least terse.

This project has no wish to speculate about sentient infrastructure. It does not have yet another vision of smartgrid to pitch. It merely aims to unpack locality and in doing so to emphasize inhabitable scale. Surely the "destabilization, deregulation, decentralization" of the electricity industry that began well before internet has been amplified and accelerated. Despite a legacy of bias toward larger technological systems, today's innovation moves toward a preponderance of the small. Millions more grid devices interconnect. Whether isolated or aggregate, and whether extending or disrupting the existing grid, these diverse elements assume more local scale, and can operate in more direct integration with their physical environment. Today's energy network finds more emphasis on intermediate, inhabitable scales of building, district, and neighborhood. There it becomes subject matter for interaction design, architecture, and urbanism. After all, there are people out there ready to call entire cities smart.

Total or Local?

Given all such background context, the question of locality may make more sense. It can help to bring built environments into the conversation. As the opening argument on islands observed, no city is a continuum, and any city needs a balance of aggregation and separation. Otherwise totalities tend to feel dystopian.

Fortunately, there are some physical limits to energy totality itself. Note some popular misconceptions on so-called Energy 2.0. Since the internet and likewise the electric grid is each, separately, a vast, vital, mostly unseen network, a lot of people outside the field could superficially imagine similarities, and even some kind of union, between the two, as if an internet of energy. Although "internet of energy" yields over a million search results, electricity itself does not just become one more role of the internet. Nor is "the" grid really a single network like "the" internet, as instead it operates separate pools.[35] To repeat and marvel, at the very moment that electrons are generated, they all go everywhere in a pool and must get used somewhere, instantly. Carefully controlled transmissions between pools help balance the larger regional differentials of supply and demand, and hence spot pricing, such as when individual power plants come up or go down, and this has long been the basis of grid control systems. Having the widest, most diverse networks really matters for power reliability and quality. Promises to deliver electricity do get bundled, dispatched, and priced in a real-time virtual marketplace, and adjustable demands for using that electricity increasingly respond algorithmically to such varying conditions. Still, it is worth clarifying that physical power itself cannot be packet switched and sent to specific addresses like data.

Totality also makes less sense under bottom-up two-way distribution. With ever more electricity being generated everywhere, clusters make more sense than universal, one-way hierarchies. When not only generation but also operations move out to the edges, then local ownership, understandings, and practices matter more. Usability arises from local information systems good enough not to require full-time monitoring from costly control rooms. It may not be the set-and-forget simplicity that so many individual consumer apps in so many other fields now tout, but it does allow smaller organizations and communities to manage on-site systems. This more abstract new layer of software systems enables the more visible improvements in solar panels and batteries. All this brings electricity back into a prominence in the everyday cultural landscape.

Physically, wind and solar generation are in view, and solar is often integrated with other construction. Whether or not anyone cares to call it smart, green building applies sensor and actuator systems that couple well with such local energy operations. Neighborhood- and campus-scale work well for heat-plus-energy cogeneration, solar power aggregation, and interconnections with water, waste, outdoor lighting, or public transit. Integrative ecological design has become a desirable local distinction.

That prominence helps recall doubts about totalities. In the consensual definition of smartgrid cited earlier, note the use of the word "every." Here is one of the worst category errors of the times: whether about a network infrastructure, a physical place, a private business, or an individual citizen, not "every" thing can or should be documented. Nor, of course, can life be reduced to only what is measurable. Nor would it be wise to let any one entity own all the documents and measurements.

"The smart city presumes to an objectivity," urban computing pioneer Adam Greenfield accused early in the 2010s, as the hype was accelerating. "The smart city is built on a proprietary platform."[36] The top-down view prevails. For example, note the popular, almost obligatory use of modular isometric diagrams. In easy self-caricature, these suggest a view of the city as if made of interoperable modules like a SimCity game (figure 3.6).

By contrast, as expressed in an influential early work (1988) by Carolyn Marvin, science technology society questions instrumentality. To Marvin, the field "introduces issues that may be overlooked when the social history of these media is framed exclusively by the instrument-centric

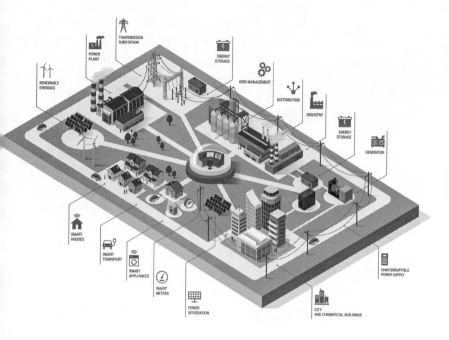

3.6   For selling the smart city, the favored visual idiom is modular and isometric, like SimCity. Dreamstime/elanabl.

perspective that governs its conventional starting point." Instruments may be easier to build and sell, but there also needs to be social practice change, often first. For as Marvin cautioned, "Changes in the speed, capacity, and performance of [devices] tell us little about these questions. At best, they provide a cover of technical meanings beneath which social meanings can elaborate themselves undisturbed."[37] To compensate, social practice research attends to what sociologist Elizabeth Shove has called "infrastructures-in-use" and "what infrastructures are for." This is not about behavior so much as intent. This is not just the invisible hand of the market for new technologies. Institutionalized practices (often unofficially so) embody "complicated relations between infrastructures and the patterns of consumption and practice they sustain."[38] Good design recognizes and affords shifts in what is considered normal.

Here, then, is an important foundation too often overlooked by the proponents of technological totalities. Social practice theory insists on the significance of local physical context. Grid awareness needs some experiential perspective. Complementary to the role of the proprietary infrastructure builders, this more open discovery process seeks the kinds of designed local interventions that trigger (or reflect) new thinking, new organizations, or a renewed valuation of place.

Science, Technology, Society, . . . and the White Paper

As noted in the opening, science, technology, and society observes how any technological system always has a cultural position. Although that stance began (thirty years ago) in reaction to modernity's misplaced technological determinism, today it also needs qualification amid the widespread dismissal of science by fossil fuel interests. Here is an important distinction. When politicians point out that science has cultural bias, generally they do so by taking *less* into account, usually by ignoring facts that don't fit their worldview. They fail to recognize how science is all about challenging old theories on the basis new findings. Science, nevertheless, does have a bias toward what is most readily quantifiable. Science, technology, and society points out that cultural position by taking *more* into account. Today when externalities, complexities, and such unmeasurables as hope and resilience play such necessary roles, and when many former certainties about systems have destabilized and now demand more heterogeneous engineering, any more useful forum thus needs more widely open scope.

Consider that stance with respect to smartgrid. Alas, despite the need for many more voices on what have obviously become complex cultural challenges, smartgrid debate suffers from what some critics call "solutionism."[39] When experts begin too many conversations by claiming to know the answer, that makes creative discoveries and social participation more difficult. However policy characterizes organizational goals, engineering solves well-bounded problems, or finance vets the feasibility of proposals, nevertheless a solutionism assumes that all these approaches can succeed by mainly instrumental, technocratic means—ones that take only the most measurable factors into account. It assumes that social or cultural prerequisites can be expressed as market demand, and that the social or cultural consequences will be good.

Under this outlook, the most expressive channel has become the "white paper," a pragmatic editorial document in which an expert organization explains its approach to a complex issue, usually to increase nonspecialist support for a proposal. Neither sales pitch, nor market analysis, nor policy declaration, but not without vested interest in arriving at each of those, a well-executed white paper puts an agenda on the table with just enough depth to engage more participants. Like it or not, any field with a complex mix of technology, policy, community, and experience now floods with white papers, far in excess of its conventional academic literature. At this writing, a Google search for "white paper" plus "smartgrid" yields over a quarter million results.

How do universities respond to this genre? For to judge by recent proceedings from leading associations in science, technology, and society studies, recent academic literature has made too little attempt to interpret this cascade of white papers on smartgrid.[40] Thirty years ago, cultural studies of electric power systems demonstrated a useful academic method on the interplay between social and technical change. This in turn became influential in the ethnography of technology, which in turn increased the emphasis on the cultural context of usability, which in turn brought the built environment back into consideration (figure 3.7). That is the history of ideas that this project is coming from.

Today such approaches have become much more widely understood. Notions of technological paradigms, frames, and momentum that once took deeply academic critiques to legitimize now arise casually in many more contexts. In what some strategists call triangulation, it is better to look for transformative change where different, incomplete orderings intersect than under a single, more complete ordering, or else in the absence of much ordering at all. Today the hyperlinked, instantaneous, saturating, and distracting character of an internet age has made such effects much more usual.

1980s academic historians:
social history of technological systems

1990s silicon valley researchers:
enthnography of technology

2000 millennial dotcoms:
contextual design for usability

2000s locative media movement:
situated technologies

early 2010s internet of things:
tangible and embedded interactions

late 2010s grid edge boom:
responsive architecture

3.7   A time line of ideas behind this project.

A New Grid Awareness

What any of that means is going to take a grand social project to enact. Electricity's latest era has become so much more than an engineering and policy challenge. Cultural values may shift. For instance, not everything concerns the price efficiencies that built the twentieth-century grid. At least some efficiencies might be of resources rather than prices. At least some resilience matters more than yet-more efficiency.[41] Participation, so important in emergences, might also matter when monitoring ever more complex systems, and for both of these reasons, in some cases cultural values may shift away from total automation. Likewise, experiential uniformity gives way to desirable contrasts in the pursuits of comfort. Design takes other physical systems into account and values more tangible interactions at an understandable scale.

To repeat in summary: in contrast to the twentieth-century legacy of centralization, standardization, and hence cultural closure about what constitutes "the" grid, today brings a period of cultural reopening. The consensus was broken forty years ago, and the flexibility of narratives as well as competition of visions and versions has been accelerating ever since. By now these many developments have reentered some wider cultural awareness. By now they indicate a new era, beyond the long and stable dominance of twentieth-century electric power utilities. Since power is such a bedrock industry to almost all others, this is controversial, potentially volatile, and worth the consideration of many more people, organizations, and disciplines.

In this hybrid forum, the internet-centric vision of smartgrid is but one narrative among many, and not an inevitable technological determinism. Likewise a more general, pragmatic project of *grid modernization* is just one narrative among many—one that is quick to assume that the fundamental arrangements of the last century need not change. New priorities of resilience suggest inevitable change. If the identity of districts, neighborhoods, and well-made buildings is to play a greater role, that brings architects and urbanists into the narrative. Since so many of today's cultural shifts and revaluations arise through apps, analytics, monitoring, and sharing, how are those better embedded into material social practices, and how do those, more so than any external rational smartgrid, now alter the cultural appreciation of electricity?

Forty years in the making, this expanding arena of cultural question can no longer be left to the managerial class of engineers, economists, and policy analysts who for so long had full control of the grid. Of course their expertise, priorities, rules, formulas, and simulations still have more accuracy and influence than any other narratives of these topics. Yet no longer are those in consensus among themselves, no longer can they afford to leave so many factors of physical space and place out of the account, and no longer do their operations seem well aligned with social, cultural, and environmental conditions on the ground.[42] Increasing the shareholder value of investor-owned public utilities no longer aligns so well with maintaining any sort of public good. Distrust of the grid, which still remembers Enron, now resides in a online world of conspiracy theories (figure 3.8). So this is an anthropological narrative too. Quite often the anthropologists have important things to say. As do historians, urbanists, systems ecologists, interaction designers, and communitarians.

To repeat for emphasis, altogether it has become quite a new era. Everybody and nobody has the answers. The status quo cannot persist. New kinds of haves and have-nots now arise, all under threat of unforeseeable catastrophe.

3.8 Distrust of smartgrid: movie poster from an award-winning populist documentary, *Take Back Your Power*. Courtesy of TakeBackYourPower.org.

Apocalyptic sensibilities rarely motivate, but neither does utopianism. Smartgrid may not be the technoidealist, universal project its stakeholders imagine it is. Local electricity debates invite a deeper sense of history, cultural assumptions, and place. Yet the prospect for local electricity portends new kinds of private opulence and public squalor. It might all seem like a mess, especially to the grid operators, or especially against desperate political opposition. But it need not result in apocalypse, and may not result in utopia. It might need a new name. Remarkably, at this writing, no other uses of this phrase come up in a Google search. Call it *smart green blues*.[43]

4 MICROGRID INSTITUTIONS

A new grid awareness began after Superstorm Sandy (2012) (figure 4.1), and then jumped after the three major US hurricanes of 2107.[1] By most estimates, Hurricane Harvey took down nearly ten gigawatts of power generation. Irma left somewhere around ten million people (and two-thirds of the area of Florida) without power. Hurricane Maria left much of Puerto Rico without power for months on end. Amid the humanitarian crisis there, in which the absence of power exposed still-deeper issues, many eyes turned toward local electricity futures. "We want microgrids everywhere," governor Ricardo Rosselló said about his Energy 2.0 plan. In this he was hardly alone.

The case of Puerto Rico demands mention not only for the harsh reality of extended power outages but also for suggesting the importance of better local institutions. Here is a short version. It was by far the largest blackout in American history, and in world history, second only to Typhoon Haiyan in the Philippines (2013), as measured by the cumulative number of electricity customer hours lost, in which it eventually surpassed Superstorm Sandy by nearly a factor of five. With over three-quarters of the lines down, the rebuilding process was remarkably slow. By some estimates,

4.1 A now-canonical image of darkened lower Manhattan amid Superstorm Sandy. Wikimedia Commons.

the number of line workers who came in to rebuild in the first month was lower by a factor of four than in Texas after Harvey the previous year, and by a factor of twelve after Irma in Florida in the same hurricane season.[2] Besides the few corrupt contractors exposed in mainstream news, poor logistics and financial planning were a constraint throughout. In a less-known side of the story, the main public utility company was in bankruptcy anyway, long before the storm.

By contrast, as the prominent cultural critic Naomi Klein reported in a long read in spring 2018, the island was facing a technoutopian influx of entrepreneurs seeking to build their own private energy islands amid the chaos and major tax havens of the physical island.[3] The irony was worsened by how many of these new "Puertopians" were newly rich from Bitcoin mining, one of the most energy-intensive activities in existence. This in a place where elsewhere people were getting by with diesel generators for a couple hours each evening, except where those generators were being stolen.[4]

Meanwhile, despite tropical sun and towns full of flat roofs, Puerto Rico came into this with less than 3 percent renewable energy. Then the storm destroyed these few photovoltaic arrays just like everything else. So despite the governor's fine wish for microgrids everywhere, it would have to be a long-term evolution from the center, and not an immediate new reality from the ground up. The pressure to rebuild as quickly as possible discouraged any new arrangements. For instance, according to Peter Fox-Penner, a respected strategist, "The only logical way for Puerto Rico— and every other storm-prone electric system—to become a series of resilient and clean microgrids is to first get the entire grid functioning and then to create sections that can separate themselves and operate independently when trouble hits."[5] For instance, Sonnen, an infrastructure builder with ample success in Germany, was building a few microgrid prototypes in Puerto Rico. Amid a brief but disturbing return to a universal outage after an accident and chain reaction in April 2018, happily these few microgrids were still up. Yet to get many more like this clearly would take new institutions, and that would take time. As Fox-Penner recommended, the effort would need new patterns for the "financing, ownership, operation and maintenance of the systems."[6] The advocacy group Resilient Power Puerto Rico likewise set a clear eventual goal for microgrid institution: "Our vision: A Puerto Rico with redundant, reflexive, and inclusive built and social infrastructures where communities across our Islands autonomously adapt to build a sustainable and equitable society."[7]

A Future More Local

That ideal belongs to a larger worldwide trend. It is beyond the scope of this writing, but plans for the approximately billion people who never had electricity in the first place have crossed into a new era. In a 2018 session at the United Nations, Bloomberg New Energies Fund forecasted that by the mid-2020s, more than half of the world's new electrification will not need grid connection, and at least in frontier areas, already the cost of independent local electricity is lower than conventional grid connection.[8] While it has become a cliché to anticipate any market's "smartphone moment," that metaphor does especially make sense in places where cellular networks were the first networks, since there were never any landlines there before them.

Meanwhile, in the wealthier parts of the world, priorities were shifting too, especially toward grid-competitive renewable energy. As identified in the opening basis for this writing, several practical trends have converged in decentralization, decarbonization, and digital technology. Despite conditions in Washington, the timing seemed right. This became so not only in familiar residential solar but also now at larger commercial and industrial scales. As just explored, the utilities, which are still responsible for keeping it all running, are having a hard time refuting these expectations. Led instead by such unlikely counterparts as Walmart and Google, by 2018 close to half of the Fortune 500 companies (hardly the usual image of environmental innovators), had set real targets in clean energy.[9] As the billion-dollar disasters kept on coming, resilience was gaining economic representation too. When at last you can get both resource economy and price economy together, and when at last resilience has both economic valuation and cultural attention, then the culture tips quickly.

When people see the lines down from storms, but new energy sources still working, right there in view, they do tend to imagine the future is local. "Power is going local," even the century-old giant General Electric declared in a media campaign that was part advertising, part public education, and as the company slashed its workforces and dividends, perhaps part defeat. Local food and music matter culturally. Supporting independent local business is almost always a good idea. Now that sensibility can include electricity as well. Much as food has its "locavores," so electricity now has its "locavolts."[10] While less likely than a farmers market to get you out on the square on a Saturday morning, local electricity nevertheless also now invites identifying with a place. It can even bring jobs and revenue, but the point here is cultural. And if ever the central grid should go down, then this local identity becomes quite distinct.

So you don't have to be an apocalyptic prepper to see the increasing benefits of more local electricity. You don't have to be a climate justice warrior to connect the future of decarbonization with the future of democracy.[11] You don't have to be a policy analyst to expect regulatory reform. By now so many immediately positive, practical aims have arisen, almost always at local scale, that this new grid awareness has found a name. For a single word to characterize these new aims, the mid-2010s brought a wave of enthusiasm for microgrids.

The Word Itself

To resume from the earlier introduction, this sticky new word *microgrid* needs unpacking. What should it mean? What are its defining local characteristics? What larger cultural wish does its popularity represent?

To repeat the most usual definition, *microgrid* means a local energy network that can run independently of the larger interconnecting grid, sometimes or all the time, by means of not only local power generation but also local network control. In the obligatory official definition from the US Department of Energy, "A microgrid is a group of interconnected loads and distributed energy resources within clearly defined electrical boundaries that acts as a single controllable entity with respect to the grid. A microgrid can connect and disconnect from the grid to enable it to operate in both grid-connected or island-mode."[12]

This has been conceived in engineering. Note that whatever its eventual cultural implications, this is hardly stated from a cultural perspective. Many more kinds of interpretations thus seem warranted.

Thus the customary proof of concept is a technical feat. Any history of the microgrid must mention the remote Mojave Desert town of Borrego Springs. Surrounded by the huge Anza-Borrego Desert State Park, with ample sun, intense need for cooling, and only a single power line to the nearest major city (San Diego), this town was an obvious candidate for its own island network. After the California electricity crisis of 2000–2001, and with a consortium of technology sponsors from 2005, the utility built what many now cite as the first diversified microgrid: not all one campus, owner, or type of use, it is able to stay up when cut off by a major storm. That came 2013, the year after Sandy. Then in another first, having integrated a nearby solar farm, in 2015 Borrego Springs ran a planned disconnect that demonstrated the all-solar powering of an entire town.[13] So in many ways, the word took off there.

Looking further back, engineers often trace the word "microgrid" to the Consortium for Electric Reliability Technology Solutions, a research group at Lawrence Berkeley National Laboratory, which began using it around the millennium.[14] A word search on LexisNexis reveals no related uses prior to then. (In one rare anecdote from 1986, Microgrid was the brand name of a graphics tablet.) The first use related to electric power appears in a *Mercury News* (San Jose) story on photovoltaics in August 2000.[15]

When the Energy Independence and Security Act of 2007 recognized a fraying national grid, federally funded research accelerated. The US Department of Energy soon recognized microgrids as a vital component. Berkeley Lab, home of the aforementioned consortium, advanced a "microgrid concept" together with a "distributed energy resources customer adoption model" (see why acronyms soon take over).[16] As the latter became a widely used template for feasibility modeling, the microgrid became a more widely operable idea.

Thus after a few occasional appearances, the word caught on in the late aughts. As measured in a word count from news, law, and policy items (figure 4.2), the use of the word "microgrid" increased tenfold in five years, 2007–2012, and tenfold again (by a factor of one hundred over the decade) from 2012 to 2017.

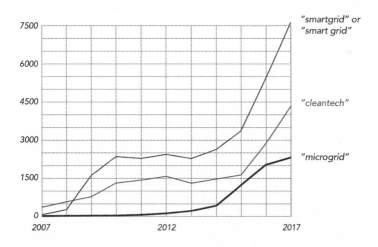

4.2 Word count of news stories on LexisNexis, 2007–2017.

The word also caught on as Silicon Valley began doing *smartgrid*. There it assumed the usual disruption ethos. For instance, in the much-cited essay "Why the Microgrid Could Be the Answer to Our Energy Crisis" (2009), *Fast Company* saw the trend: "In the end, the visionaries of the microgrid are confident that with the help of evolving technologies and dropping prices, the unstoppable force of the American consumer will steamroll the last immovable industry out of the way."[17]

Within five years, this counterforce had become a recognizable market phenomenon. "Microgrids are poised to become an integral part of North America's energy transformation," declared a collaboration of the major infrastructure builder Schneider Electric with the nonprofit International District Energy Association in 2014. "It is a coming together of several societal, market and technological trends and changes."[18]

Always an example in journalism, when in that year the *Economist* did its part to bring the "microgrid" into the mainstream, it was characteristically pithy about the word:

America's electricity grid is a mind-boggling mess. . . . America's power industry talks about creating a "smart grid, a digitally connected network, automatically monitored and balanced, to solve the problems." Yet this misses the main challenge. [Smart grid] has little effect on the grid's overall ability to handle weather-related onslaughts that hit wide areas or a big cyber-attack. . . . Microgrids, with their own electricity-generating capacity, are a better bet.[19]

"Local [electricity] is the critical missing piece in our outdated, centralized electrical system," observed the Clean Coalition, a public education and policy organization, at the launch of its ongoing Community Microgrid Initiative. For this widely watched effort, the Clean Coalition has provided

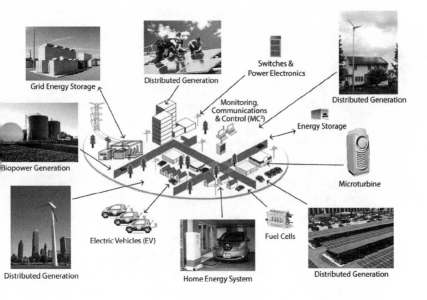

Grid Energy Storage

Distributed Generation

Switches &
Power Electronics

Distributed Generation

Monitoring,
Communications
& Control (MC²)

Energy Storage

Biopower Generation

Microturbine

Electric Vehicles (EV)

Fuel Cells

Distributed Generation

Home Energy System

Distributed Generation

4.3   "Elements of a Community Microgrid," in what is probably the single most sampled isometric diagram in the field. Courtesy of the Clean Coalition.

the one most sampled, reused, and by now iconic instances of the obligatory modular isometric diagram (figure 4.3).[20]

The Institute for Local Self-Reliance, a community enterprise resource at work since the 1970s, declared electricity's latest era. "The electric grid is no longer a 20th-century, one-way system. A constellation of distributed energy technologies is paving the way for 'microgrids,' a combination of smart electric devices, power generation, and storage resources, connected to one or many loads, that can connect and disconnect from the grid at will."[21] The institute's oft-cited "Mighty Microgrids" report (2015) saw a new field ready to explode as soon as policy barriers to community scale were removed.

For metrics of a boom, the prominent market forecaster Navigant Research appears to be the most cited source. In an annual forecast in 2014 by Navigant Research, Peter Asmus, a highly respected analyst in the field, wrote, "Today the industry is moving into the next phase of project development, focusing on how to develop projects on fully commercial terms. . . . Navigant Research expects annual [US] microgrid capacity to increase from 685 MW in 2013 to more than 4 GW by 2020, a compound annual growth rate of 29.3 percent.[22] Two years later, in 2016, however, Asmus cautioned, "The hype cycle on microgrids appears to have hit the crescendo level, causing at least one commentator to say 'microgrids are the new kale.' This, of course, refers to the trendy vegetable alternative to lettuce and other leafy greens."[23]

When the growth curves are steepest, often that is because the adapters are still few, and the overall volumes are still low. As of 2018, when to much celebration the overall amount of renewable generation in the national portfolio rose past 15 percent, nevertheless the amount of overall electrical capacity managed via microgrids was still well under 1 percent.

Yet in that same year, Navigant's trackers identified as much as twenty-four gigawatts of microgrids in the works worldwide, with the United States meeting earlier estimates, and "energy storage for microgrids" a particularly hot market. Although US numbers were not shown in its public reports summaries, Navigant's ten-year global forecasts for 2017–2026 anticipated at least a tenfold capacity increase in that, and of potential relevance to architecture's grid edge, more like a twentyfold increase in "commercial and industrial building to grid integration." For the more inclusive category of "smart city solutions and services," Navigant's ten-year forecast was "to grow from $40.1 billion in 2017 to $94.2 billion by 2026."[24]

By any measure, underinvestment in US infrastructure creates potential for countless specialty infrastructure-building markets to arise. By the end of the 2010s, major grid builders like Schneider and Siemens were admitting that the microgrid market had been underestimated, and were forming many new partnerships and ventures in this fast-growing field. Many urban campus and redevelopment district sites had plans of their own. Likewise, city governments (New York's in particular) were advancing projects and policies that suggest an enduring boom and no mere speculation. Look to faster network media for the latest on the story, perhaps, but do also imagine that the kind of long view fostered by science, technology, and society may help understand the dynamics of choices ahead, and did, in this case, see something coming.

What It Is and Is Not

The rise of this word has become a catalyst for new grid awareness. Although working microgrids remain expensive and few, already some cultural retrospect on the late 2010s shows an idea that is here to stay. Yet soon an appealing term means many different things to many different people. For as science, technology, and society work has emphasized, even where meanings eventually settle, the paths by which the culture gets there are seldom linear. Today this hot new term deserves many new perspectives, and at least some of those with an emphasis on local place, and thus at least sometimes on the built environment.

More disclaimer seems necessary here. This writing is not an argument for the immediate practicality of microgrids. That would take more technical knowledge or policy skills than almost anyone, much less an outside observer, seems to have at the moment. This is not yet another amateur manifesto for a bright green future. This is not a pitch for the advantages of microgrids. Instead, here is simply an argument that the distinct rise in talk of microgrids reflects a cultural need for locality and autonomy that has not been getting met by the electric power industry. In particular, here is an argument for locality amid *the smart city* that has not been much explored for its relevance in architecture and urbanism.

With that qualification of approach in mind, consider a bit more detail. Since the web is full of all the cool things that a microgrid could mean, here begin by acknowledging what a microgrid is *not*.

A microgrid is not a single source and application of energy. For example, a direct current, solar-powered LED lighting system should instead more accurately be called a *nanogrid*. Direct current works well for short distances, can power electronics, has relatively low physical losses, and saves on costly alternating current inverters.[25] Incidentally, that is not such a new idea itself. The popular self-contained Delco-Light Plant (figure 4.4) that was introduced in 1916 was a primitive direct current nanogrid of a sort. Tens of thousands were sold, and working hierlooms are apparently popular among preppers today.

To understand this distinction, note that today a microgrid might connect several nanogrids. Hence the number of grids on "the" grid could be in the millions. For the language used by its advocates, this helps explain the popularity of galaxy metaphors. For understanding the term's relevance at district or building scale, it gives a name to a very practical new hierarchy: as their numbers increase, location-specific microgrids help manage application-specific nanogrids (figure 4.5).[26] A microgrid is not just another layer of control systems.

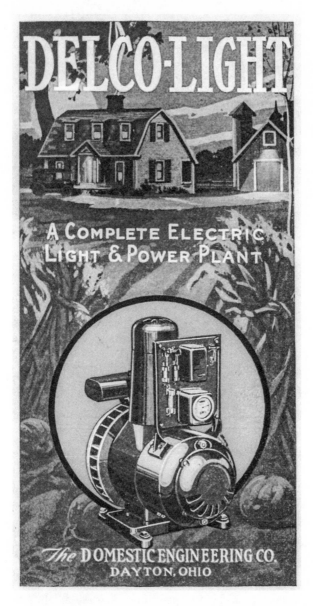

4.4   Century-old nanogrid technology: a 1916 advertisement
for the Delco-Light Plant. Courtesy of Duke University Libraries.

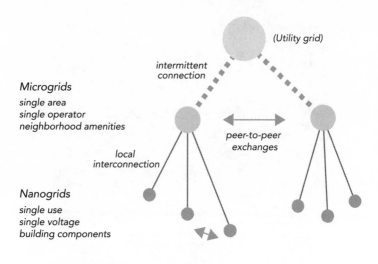

Microgrids

single area
single operator
neighborhood amenities

intermittent connection

(Utility grid)

peer-to-peer exchanges

local interconnection

Nanogrids

single use
single voltage
building components

4.5 Microgrids for locations can manage nanogrids for applications.

Where local distribution networks, which are already part of the legacy grid, now get "modernized" for resilience and distributed generation, they are better called *milligrids*.[27] So far that is a less widely used term, and one unlikely to gain as much local identity. By contrast, a microgrid has a contiguous, bounded area, is likely to have a distinct owner or constituency, capable of operating as an independent island, and separated from the main grid by what is often just a single connection point.

A microgrid is not just a diesel backup generator. Most microgrids integrate multiple uses and sources, and preferably some storage, all under a new degree of local network control. The more interesting ones do not just provide isolated backup and do not just run on noisy, stinky diesel.

A microgrid is not just any site's solar-plus-storage scheme, although in many cases it would be better off becoming one.[28] So far, most rooftop solar installations have depended on the grid for conversion to reliable, high-quality (that is, steady voltage and frequency) power. Only recently have the necessary technologies (inverters, storage devices, quality regulators, and load dispatchers) become affordable, usable, and allowable for larger numbers of small users. Among these, storage has become the most speculative technological quest at the moment.[29] A community solar initiative with all these would indeed constitute a microgrid. Without them, however, a solar rooftop panel powers nothing when the grid goes down. That simple fact does more than any other to explain a microgrid boom.

Similarly, a microgrid is not just any aggregate of distributed small producers. That is more a market policy phenomenon, albeit one with increasingly familiar jargon. Parties that credit excess generation back to the grid under "net metering" or sell it via separately metered "feed in tariffs" are sometimes known a "prosumers." Since it is useful to aggregate such distributed diverse sources into entities large enough to play in wholesale energy markets, there is good business in organizing "virtual power plants." Although clusters of microgrids make it easier to do them, so far those remain mostly separate and complementary entities. (But to repeat, although some people conflate the internet and energy as if one big cloud, data can be packet switched and addressed, but electric power cannot, only the promises to deliver it.)

A microgrid is not (yet) much of a data platform. That is certainly an interesting prospect for new social patterns, but so far one without prominent prototypes. Although data harvesting drives the economic feasibility of ever more things today, and already has clear leaders in energy user data aggregation for instance, as yet there is no dominant model for on site local electricity platforms. The technical challenge is recognizable, for as the number of devices and organizations at *grid edge* proliferates, the amount of data must soon overwhelm any central organization. This also raises social and organizational questions. Which data are valuable to whom?[30] Where data from, say, direct current nanogrids accumulate locally and are of little interest elsewhere, what kind of platform will manage that? Even just for a small set of buildings, that invites some new kind of services or organizations. While this argument points toward the prospect, and campus sites have obvious advantages, a microgrid is not yet an identifying feature of cultural institutions, nor itself a new kind of infrastructural institution.

Last but most basic, a microgrid is not affordable just anywhere. It adds costly wiring, equipment, and procedures to a resource that is already available at a very low cost. To install one when cheap, reliable electricity is already available almost everywhere thus appears as a luxury. Those have their role; America goes about many things overequipped. In the competition for workers, for example, the best sites want all the latest features. But the point is to find some practical, affordable value to resilience. Only when the prospects of systems outages and local resilience get monetized do microgrids seem more feasible. After all, since spending on devices has so long grown faster than spending on electric service or infrastructure, the implicit value of reliable electricity has constantly been increasing. It is difficult to put a price on dependency and, likewise, disaster risk. So the better bet appears to be on the everyday workings as resilient clusters. That is where this writing is heading: What is the value of archipelagos?

To find the consensual wisdom on any of these themes takes some keeping up with the news.[31] For the general reader now jumping in, two highly respected writers especially deserve mention. David Roberts, one of the most widely read environmental journalists of the times, was at work on a microgrid book of his own. Long known through the green magazine Grist, and now in a front-page role at mainstream Vox, Roberts has run excellent overview stories like "Meet the Microgrid, the Technology Poised to Transform Electricity."[32] Across a series of brightly diagrammed posts, he especially explains the microgrid in the context of a much-larger agenda to "electrify everything."[33] Besides benefits from the rise of renewables, there are two important but less recognized sides to that goal: for users, decarbonizing the grid makes everything on it cleaner too; for utilities, much of this new load on the grid would be controllable, which would help offset increases in uncontrollable volumes of wind and solar generation.

For an especially readable voice from the industry, the most consistent interpreter across these years has been the energy journalist Elisa Wood, whose *Microgrid Knowledge* (and conference) has aggregated more stories better than any other feed of its time. Wood has chronicled the rise of the microgrid concept from a niche and a long bet toward a viable play for the majors of electrical infrastructure—(Schneider and Siemens in particular), and by now an emerging market in which (unlike some previous green technology trends) North America is a leader. Whether on the superstorms, the solar-plus-storage boom, the policy debates, or increasingly, the public good, her news feed not only belongs in any *hybrid forum* but indeed also convenes some of them. To Wood, this is clearly not just a technical domain. Indeed in a 2016 survey of over two hundred major players in this fast-developing field, "only 10 percent named technology as a hindrance to microgrid growth."[34] In 2018, Wood could declare microgrids as an agenda "for the greater good,"not just ostentatious backups for the wealthy few, but a basis for community resilience, with services back to the larger grid.[35]

As the California Public Utility Commission has observed early on (2014), "There is a national interest in successful development of microgrids." Here is a respectable claim that the utilities cannot shape this alone. "The question is not does the industry encourage or discourage this development. . . . This development is happening whether the utility, or regulator, encourages it or not." With a national interest perhaps complementary to the one that first universalized the grid in the 1930s, now came a prospect that "upends the traditional means of providing and consuming electricity . . . [in which] widespread and successful implementations of microgrids will upset the century-long view of 'the electricity grid.'"[36]

New City Islands

By now a few favorite cases demonstrate more than engineering proof of concept. Here begins a cultural phenomenon that no print retrospect can cover quickly enough. Now the usual notions of a top-down, all-connecting smart city encounter something quite different. Whatever else all the above properties of microgrids have come to mean, in urbanism they suggest new kinds of identity for districts and institutions.

While it might indeed feel like the new kale to read too much into these, it would be a mistake to treat them solely as engineering. Consider a few pilot cases from these recent years. At this writing many more microgrids are in planning than already built. So these examples may soon seem overexposed, or dated, but also historic and influential.

In a setting with a density quite opposite that of a Mojave Desert town, a new microgrid prototype has been arising in Chicago's historic Bronzeville neighborhood. In its community choice policies, renewables targets, and high investments in grid edge technologies, by the late 2010s

Illinois had become one of the most innovative states outside California. Amaren, a leading public utility there, was among the first to pilot a higher-voltage microgrid within its distribution network. Commonwealth Edison (already mentioned in other contexts as a pioneer then and now) was building a multiowner microgrid in partnership with the Illinois Institute of Technology. Given that Bronzeville is such a significant African American historic site, many public commentators sought stronger communitarian roles for energy democracy. Others wished for any practical new backups to go first to supermarkets in the urban food desert. And of course, plenty of citizens expressed disappointment at the use of fossil fuels. Yet Commonwealth Edison and the Illinois Institute of Technology mostly wanted to demonstrate intermittent interoperation between two islands. The solar-plus-storage instances would appear soon enough (and in 2018 prototypes were underway at the area's Dearborn Homes). This multi-owner project is widely watched because it is an important model for future resilience strategies.

In a setting with much higher wealth and density, a microgrid serves the largest private high-rise development project in New York City since Rockefeller Center: the riverfront high-rise district at Hudson Yards (figure 4.6). In preparation for the next Sandy, it provides local electricity from heat-plus-energy cogeneration, via a campus water plant supplemented by gas-fired microturbines in the individual towers. At thirteen megawatts overall, this is one of the larger such projects in the city.[37] In the formerly uniform and universal service of electric power, this project demonstrates new appeal and advantage. In a workplace market where the competition for talent now emphasizes high-performance infrastructures, on-site work-life balance, and more creatively engineered environments, Hudson Yards demonstrates prospects for standing out. "We want

HUDSON YARDS™
NEW YORK

ᴇNGINEERED ᴄITY

Hudson Yards will be far more than a collection of tall towers and open spaces. It will be a model for the 21st century urban experience; an unprecedented integration of buildings, streets, parks, utilities and public spaces that will combine to form a connected, responsive, clean, reliable and efficient neighborhood.

CONNECTED NEIGHBORHOOD

Communications will be supported by a fiber loop, designed to optimize data speed and service continuity for rooftop communications, as well as mobile, cellular and two-way radio communications. This will allow continuous access via wired and wireless broadband performance from any device at any on-site location. We're as good as future-proofed.

 Digital antennae service (DAS) for cellular and two-way radio   Rooftop satellite   Wireless responders   ⅢⅢⅢⅢⅢ Fiber Loop

RESPONSIVE NEIGHBORHOOD

Hudson Yards will harness big data to innovate, optimize, enhance and personalize the employee, resident and visitor experience. Supported by an advanced technology platform, operations managers will be able to monitor and react to traffic patterns, air quality, power demands, temperature and pedestrian flow to create the most efficiently navigated and environmentally attuned neighborhood in New York.

 Building data-capture sensors (systems, equipment)   Electrical and thermal sub-metering   Environmental sensors (air, noise, other environmental factors)   Advanced technology platform

CLEAN + RESPONSIBLE NEIGHBORHOOD

Progressive cities are moving toward organic waste separation systems to reduce landfill costs, methane emissions and greenhouse gas emissions. Hudson Yards makes organic waste collection convenient and space efficient by utilizing grinders and dehydrators to reduce food-service waste to 20% of its initial weight and volume.

Additionally, nearly 10 million gallons of storm water will be collected per year from building roofs and public plazas, then filtered and reused in mechanical and irrigation systems to conserve potable water for drinking and reducing stress on New York's sewer system.

 Organic-waste disposal system   Stormwater Tank

RELIABLE + EFFICIENT NEIGHBORHOOD

Whatever the disruption—super storm, brown out—Hudson Yards will have the onsite power-generation capacity to keep basic building services, residences and restaurant refrigerators running. It doesn't hurt that being built above a rail yard means our first level is well above the flood plain.

Hudson Yards' first of its kind microgrid and two cogeration plants will save 24,000 MT of CO_2e greenhouse gases from being emitted annually (that's equal to the emissions of ~2,200 American homes or 5,100 cars) by generating electricity, hot and chilled water for the neighborhood with over twice the efficiency of conventional sources.

 14.5 megawatts of cogen   18 megawatts of Tier 4 diesel generators   Con Ed Utility Grid   Microgrid Breaker

 Hot/Chilled water plant   Hot/Chilled water line

4.6   New York, large scale: Diagram of the energy cogeneration microgrid for Hudson Yards. Hudson Yards press kit.

a large-enough system on site in case the grid goes out, so we can keep Hudson Yards sort of a beacon of New York City," the developers observed.[38] Whether this beacon will guide the less fortunate in their emergency responses, or as some critics contend, just illuminate yet another category of haves and have-nots, remains to be seen. Right now the more important point is that fancy private infrastructure makes big economic sense.

Nevertheless, small-scale ad hoc institutions also play a vital role, particularly at expanding local participation. For instance, few cities have more locavolts than Brooklyn. As the *New York Times* observed, "Brooklyn is known the world over for things small-batch and local, like designer clogs, craft bourbon and artisanal sauerkraut. Now, it is trying to add electricity to the list."

In New York, the Brooklyn microgrid is conceived to work with the conventional grid, which is in the midst of a reboot under Gov. Andrew M. Cuomo's directives to make it more flexible, resilient and economically efficient while reducing greenhouse-gas emissions. That effort, known as Reforming the Energy Vision, or REV, includes encouraging the development of microgrids and more active community participation.[39]

Brooklyn Microgrid crowdsources votes for suitable solar-generation sites on community asset sites such as churches and schools, and then approaches those sites with proposals to plan and crowdsource funds for resource development. For individual users, it also helps identify, schedule, and share green energy resources, preferably local to keep the money in a "circular economy" in the neighborhood. It helps monetize energy savings (negawatts) too. Its smartphone app "takes the meter off the wall and into your hand" (figure 4.7).[40] This is not just ironic protest. The peer-to-peer (blockchain)

4.7 New York, small scale: Brooklyn microgrid. Courtesy of Brooklyn Microgrid.

trading and cryptography in the Brooklyn Microgrid have attracted the participation of infrastructure giant Siemens and seeded the energy company Lo3.

New York's abundance of local electricity agendas illustrates how ad hoc organizations explore new prospects and eventually shape new normals, best under conditions where institutions, state programs, and reformed energy markets all shift simultaneously. It is a combination of local innovations, large-scale infrastructural aging, and proactive top-down policies that best accelerates change. At this writing, the New York statewide Reforming Energy Vision (first introduced in 2014) has become the most widely cited such arrangement in the United States, and the New York energy research prize competition has become one of the most watched local innovation competitions to help communities (and not just single-owner sites) build microgrids.[41] Ambitions are high: New York's goal is to seed market reform, do away with many energy subsidies, and identify important new kinds of institutions, particularly new kinds of distribution, many of them at neighborhood scale. According to official estimates, the state's aging infrastructure would take billions just to maintain as is—as much as $30 billion over ten years—much of it for less clean, less often used peak capacity. Alas, since peaks do increase even where overall demand has begun to decrease, the whole system is by now oversized. So while the state wants market reform and refers to "a new institution" for an energy distribution system platform, there is debate about whether the utility companies should own all of it. Green advocates want more democratic governance. Neighborhood and campus microgrids appear an important negotiation point in such democracy. Who gets to play?

Lastly here, for what may well be the single most influential case, since it is at a more usual scale and mix of development than these others, the Pecan Street Project in Austin, Texas, has developed a residential microgrid platform. Back in the early 2010s, when the city planned a sustainable neighborhood for the build-out of its former Mueller airport site, it included a smartgrid demonstration project. Because Austin controls its own electric utility, it had unusual flexibility in this. This has already become something different from the urban district energy plant or rural electrical coop of the past, and more ambitious than the prevalent community solar model of recent years. The key is the design and integration of the built environment along with its related systems of water, cooling, mobility, and neighborhood identity. The engineers and policy analysts of the electric power industry might not have come up with that on their own.

As a result, what began as a smartgrid demonstration project has led to a model for linking technology, behavior, and building. Combining the advantages of a city-run utility, a large-scale green residential development, and a culture of technological early adopters, Pecan Street has built what may be the most portable microgrid prototype so far. Its software model connects domestic energy and water usage, and collects data for both grid operations and behavioral research. The latter is now supplied through a nonprofit organization providing access to one of the largest collections of "disaggregated customer energy data."[42] Open data curation, so different from the predatory practices of utilities and virtual markets that smartgrid opponents so often fear, here invites a new kind of service organization.

Meanwhile on the ground at the Mueller site, Pecan Street has pioneered a new kind of institution, situated between public utility, city hall, and neighborhood organizations. Here neighborhood identity depends on qualities of the built environment that go beyond the usual attributes of green building. Environmental science and policy researchers Jennie Stephens, Elizabeth Wilson, and Tarla Rai Peterson, who have interviewed a range of Pecan Street participants early in the project implementation, see advantages in "citizen empowerment."[43] Having a stake in development and cultivating some everyday awareness helps people get past paranoia (or apathy) about smartgrid as some all-seeing, liberty-reducing project of state and corporate control. High among the challenges of small-scale energy is the large-scale number of players, no longer just the utility companies, that must become aware of their respective practices and each assume some responsibility. As Stephens, Wilson, and Peterson observed, "The number of actors does not decrease as the project becomes smaller."[44]

Today's sensibility of perpetual smartphone connectivity has transformed cultural expectations anyway; participation, monitoring, and measurement have come to many more resource flows in everyday life.[45] The organizers and aggregators of these information streams and shadows may assume civic roles, or at least are useful counterpoints to the command-and-control utilities. That a few pioneering microgrids like Pecan Street have gained so much attention illustrates a widespread latent recognition of the need for such pilot projects.

As if to remind of these sociopolitical prospects, a large work of public art lines the highway near the approach to the Mueller site. SunFlowers, a series of twenty-foot stems bearing seven-sided "petals" of solar collectors for the site's outdoor lighting, has become one of the more prominent local landmarks to local electricity.[46]

Districts and Institutions

By now it seems clear that district energy strategies provide the most obvious, practical context for urban microgrids. It would be difficult to investigate one without the other. As the United Nations Environmental Programme observed in a casebook from the mid-2010s, many nations have put district energy at the center of their energy strategies, and many organizations "from the Wall Street Journal to the International Energy Agency tout district energy systems as the fundamental solution and 'backbone' of the sustainable energy transition."[47] For instance, the International District Energy Association has been a prominent voice in public education, reshaping regulations, and bringing the nascent microgrid industry into larger urban designs. Since redevelopment districts become both social experiments and economic generators, and since frankly even the most totalizing technofuturisms pilot in districts and not citywide, this is the scale to watch.

As evident from these early cases above, microgrid work has best begun at the scale of the campus, whether universities, hospitals, or larger-scale housing estates (for example, Co-op City in the Bronx). With a legacy of their own heat-plus-energy cogeneration, these campus-scaled institutions have been natural first movers on microgrids. Universities in particular have several advantages: contiguous campus, sole ownership, longer-term investment strategies, higher social conscience or at least better consensus building, and an unusual schedule of use. In one prominent instance, Princeton University served as a community haven and staging area through Superstorm Sandy.[48] As a consequence, its directors now chair the district energy association's Microgrid Resources Coalition. There, as director Ted Borer has explained, more local control provides not only resilience but also everyday efficiencies, not only for a campus but in some ways grid-wide too.[49]

Today Princeton's microgrid demonstrates state-of-the-art dispatching and load-shifting relationships with the regional utility grid. Overall peaks get reduced, local pockets of higher outage risk get better managed, and related resources like heat, waste, and mobility get tied in. This is still a long way from the microgrid becoming a perceptible part of place identity, as, say, Hudson Yards aims to do, but it is also a long way from the century-old campus powerhouse image.

For an exemplary case in neighborhood aggregation, and one emanating from the urban design discipline, the Oakland EcoBlock project demonstrates the benefits of local systems integration. Designed at the scale of a single city block (approximately thirty houses), but interpreted as a prototype for neighborhood projects, redevelopment districts, municipal policy innovations, and perhaps new kinds of hyperlocal institutions, the EcoBlock project integrates waste, water, solar power generation, architectural retrofits, and community gardens. "The block-scale is considerably more efficient & cost-effective than the individual house-scale in achieving resource efficiencies, and takes advantage of emerging energy generation legislation and information systems," the project explains.[50] Block-scale biogas digesters, solar power aggregation, rainwater harvesting, and gray water garden irrigation all help justify house retrofits, which in turn repay the costs of these systems. Yet the character of the district, based on the well-made housing types, also drives the project. It matters to have a legible sense of place worth maintaining and a city block scale worth propagating.

In an early example of an anticipated archipelago, the city of Pittsburgh has advanced a plan that was among the first to interlink multiple cogeneration schemes into a cluster of microgrids. Its interconnected islands are to include not only the campuses of Pitt, Duquesne, and Carnegie Mellon universities but also the central business district and the

North Side stadium zone.[51] Driven by Department of Transportation funding, and with provisions for one of its microgrids to recharge a fleet of city-owned vehicles and the redevelopment of the long-distressed Uptown neighborhood in between, this project demonstrates the role of district energy in urban design. As the strategists of this project have set out to explore, new kinds of microgrid standards might make it easier to interlink islands than to refurbish whole the legacy centralized grid. This is speculation, but as the Institute for Local Self-Reliance has put it, "Some microgrid advocates see the U.S. grid eventually becoming largely a grid of microgrids with the central grid acting as a back-up system."[52]

Note that in larger strategies for almost any city, the long-established urban campus provides a natural counterpoint to the brand-new special redevelopment zone. Together these point toward a networked city (figure 4.8) made of something more interesting than so many separate individual customers on one large end-to-end net. As must be taken up in an argument about archipelagos to follow, these islands of deeper connectivity and identity can often do more for social practice change. If there is one thing to say about them, as the anthropologist Mary Douglas once so memorably put it, "Institutions confer identity."[53]

Outlook

According to the classic interpretation from mid-twentieth-century urbanist Kevin Lynch, memorable architectural character is just what a district needs. A district is a "set of characteristics."[54] Not just an administrative resource zone, thus a district is also a core phenomenon of psychogeography. In the famous five elements identified by Lynch, spatial mental models grow from "paths, edges, districts, nodes, and landmarks."[55] As social practice theory would be quick

4.8 Microgrid institutions in an urban archipelago.

to emphasize, good spatial mental models anchor much else about belonging, participation, and adaptation. For instance, a district is a set of habits. Its physical pattern of participation is the location of cultural values.

This is just where the built environment may matter to the microgrid boom. Habits of spatial enactment now embrace local differences, with the presence of physical components as tokens. This is certainly not just an engineering and policy problem. It goes beyond the performance of well-integrated systems. It is no mere set of handheld apps. It connects local energy to the physical form and character of the city.

Here is cultural opportunity and challenge too seldom taken up in science, technology, and society. Do not overlook the role of the built environment. As adaptive local practice drives a microgrid boom and new kinds of institutions emerge, it might be worth remembering that many an institution would be something much less without its architecture, and that architects do much of their best work for institutions.

Institutions play important roles in the cultural imagination. Whether venerable and landed, like a university, or ad hoc and virtual, like a solar power aggregator, institutions shape and reflect what their members identify with. Their identity process comes from shared goals, places, histories of participation, local protocols, and civic recognition. To identify with a new normal is to have lived with it, habitually embodied, for cultural reasons and not simply to have bought the latest technologies to install.

So far the microgrid is mostly just an interesting idea. It is not yet a major market trend, nor just a turn in state policy. Before those larger kinds of forces arise, usually a smaller number of enlightened players lead social and cultural change in what is considered viable, and even normal. That is a core assumption for this project. The smart city cannot simply be engineered from the top, legislated, and brought to market. The microgrid boom affirms how a more adaptive process works in more heterogeneous, multidirectional, bottom-up ways, in which cultural values may shift.

Arguably the most important next microgrid institution has to occur at the scale of community. Beyond the housing estate or the university campus, groups with multiple owners and diverse agendas need new organizations and rules on local electricity. For instance, as a first move, more kinds of aggregates need the right to cross public utility rights of way with wires of their own. Groups that are not chartered institutions like universities or hospitals need to organize community microgrids among multiple buildings, and across city streets. To work this out, today a galaxy of self-organized, internet-enabled, local resource organizations operates where neither the market nor the state would succeed alone. Start-up companies, chartered nonprofits, community advocacy groups, and virtual exchanges uphold unvalued resources as well as create new patterns of cultural valuation.

These are social and political choices, related to changing values about design for living. Any great transition takes social practice, not just individual consumer choice. A sense of agency arises through shared contexts, dispositions, and goals. Social practice operates in and on place. To identify with a place, through histories and habits of engagement there, is different from having its identity declared to you like a brand. To uphold the many nonmonetized aspects of place often involves nongovernmental and extramarket considerations. To identify with new kinds of organizations that help adapt places frequently comes from small shifts in social practice. This gets internalized as dispositions, or what sociologists call *habitus*.

The more that adaptation becomes the order of the day, the more important it becomes to apply these basic social principles to things formerly considered value-neutral engineering problems. Toward microgrids, this means seeing beyond the basic desire for reliable backup power and into other, more integrative aspects of resilience, and also into the social practices that accompany, adopt, and often initiate them.

Many of these occur at district scale. Location-based media already transform mobility, lodgings, food, funding, aspects of citizenship, and valuation of place. Hyperlocal media focus these developments onto well-established community, often with tangible manifestations in proximate physical space, often at street level.[56] Local participation can shape data flows and algorithms in ways that imposed proprietary platforms cannot. If a meme is to hold and to help accelerate some greater transition, it must mean something new to citizens, organizations, neighborhoods, and their designers. Local electricity has become such a boom, and microgrid has become such an urban design consideration.

Again, to build a microgrid inevitably adds a new layer of costs, responsibilities, and technical challenges to a resource that one could formerly take for granted. Like any agenda of resilience, it raises fundamental questions about what matters most. Like any debates on values, that takes rich institutional context. Soon enough the legal, financial, and organizational models need to change. Where enlightened players lead social and cultural change in what is considered viable, even normal, often they get there before the market or the state. Whether official or ad hoc, these often operate as institutions. Whether capital *I* or small *i*, institutions typically make purposeful use of good design in their mission to confer identity. To innovate "microgrid institutions" suggests not only new organizational entities but also changing social expectations.

Thus by now it seems fair to declare a cultural opportunity and challenge: local electricity now invites many more players, projects, and ways of thinking. Yes, its technologies of generation, storage, and local network operations have improved dramatically. But so too must its institutions. Local electricity invites reflectivity, writings, design speculations, and public arts.

5 ARCHITECTURE'S GRID EDGE

As many more disciplines take an interest in the microgrid boom, architecture deserves to be one of them. Most electricity gets used in buildings, after all (figure 5.1), and a good building is more than just a place to turn on the lights.[1]

As a way into that interest, start from *grid edge*. What would it mean to modify that expression for architecture? Again, grid edge describes all the millions of local, often independently owned and operated devices now attaching to centralized electricity grids. Grid edge reverses economies and roles not just in wind and solar generation but also network operations. Although many grid edge components go into utility-owned substations and distribution networks, others are built into inhabitable surroundings, where now they belong to independent networks at campus or building scale. Although some of those increase automation, others become visible, physical components, known through everyday use, and identified as part of the place. This involves not only solar panels but also batteries, direct current lighting systems, responsive facades, demand responsive equipment, and a stack of local network operations. Increasingly these communicate not only within a central utility but also within a building's own network and also on the Internet of Things.

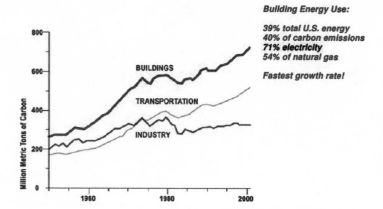

Building Energy Use:

39% total U.S. energy
40% of carbon emissions
71% electricity
54% of natural gas

Fastest growth rate!

5.1 Most electricity gets used in buildings: An oft-cited graph from the year 2000. Lawrence Berkeley National Laboratory.

Among those, ever more improve the adaptability of a site. Together they suggest a newly modified expression: *architecture's grid edge*.

Memes do proliferate; at this writing, a Google search on "smartgrid" returned approximately 7.5 million items, "smart grid" 6.3 million, "microgrid" 1.17 million, "smart building" 540,000, and "green building" 18.3 million. "Architecture's grid edge" is not one of those memes: a search for it returned zero items at this writing.[2] So for the purposes of this project, you might ask whether architecture's grid edge now becomes a thing and a topic of creative work in itself. Imagine such work as a way for architecture to engage the smartgrid meme and especially the microgrid boom. Architecture's grid edge is all the elements built into everyday inhabitable space that blur the distinction between building networks and power grids. It is all that is physically present that helps make local electricity knowable, tunable, and sometimes participatory. So if there must be a neologism to this project, let it be that.

Then to bring in one more term, architecture's grid edge consists of the *material systems* that make local electricity more tangible, knowable, and usable. Countless familiar plug-in appliances already do that best, of course; the point is that building components, and then the nano- and microgrids that they comprise, could now do better by way of their physical and spatial circumstances. In this regard, material systems belong to a more general domain of integrative design.

Since this tends to be easier to understand at object scale, consider an inhabitable object: a superlight backpacking tent. This illustrates well how material's affordances and constraints get realized in the design of a spatial configuration. It also illustrates heterogeneous engineering: it involves propositions that transcend the analytics of any one design discipline. Often that occurs as a process discovery: design improvements occur amid projects in progress, more so than from defining a problem reductively before beginning any implementation. Material systems become worth emphasizing wherever research can occur through making, and discovery through assembly.[3]

This seems especially so for things called smart. When properties of material, assembly, scale, situation, and con-figuration in effect play informational or computational roles in themselves, and when sensors, actuators, and transmitters can be built into the features of an assembly itself, a system behavior can emerge that would not be attainable or pre-dictable from physical or digital considerations alone. This is not the moment to go more deeply into this. Suffice it to say that adaptive materials and structures have become a hot topic at the intersection of material sciences, engineering, architecture, and interaction design.

Thus in any larger conceptual basis for architecture's grid edge, remember that form itself informs. That is alas sometimes forgotten amid a time oversaturated with messages, and with applied annotations, instructions, and apps for everything. Not all that informs has been sent like data packets; often an intrinsic configuration such as a footprint, a doorway, or a dynamic space frame also informs. Thus in a world of physical/digital hybrids, the idea of material systems goes beyond the assumption that information is an applied overlay. Clothing, household products, and vehicles all demonstrate as much, and now the larger built world is catching up. Augment reality, perhaps, but do not cover it in apps and instructions. These are core principles not only of architecture but also, increasingly, tangible interaction design.

While that may seem obvious for some aspects of architecture, it has not been explored much for microgrids. There is not yet a rich domain of grid edge material systems in architecture; rather, there are only a few beginnings. Note how cleantech and microgrids started as fringe speculations themselves, however, not so long ago. Now as the boom in local electricity makes grid edge into a more usual idea, it makes sense to explore how that becomes tangible, inhabitable, habitual, and hence more culturally understandable.

Intermediate Scale

It may not seem obvious, or even reasonable, to make such claims. This argument requires careful qualification. This is not a whole new architecture, as so many design enthusiasts so often love to seek and proclaim. This is not a bright green Jetsons technofuture where responsive buildings become just anything on demand, nor a dark green dystopian future where only a few well-secured compounds still have reliable power. This is not just about building automation systems, however grid-alert those have become. It is not just better mechanical systems. It is not about smart home consumer accessories. It is not just for operational efficiencies.

This argument needs some disclaimer on the latter. Yes, everyone needs to save some energy, but no, that is not interesting, and no, so much individual, app-assisted choice does not accomplish much without a change of habits and values to accompany it. Overarching change in organizations takes, in the larger sense of this word, new architectures. Otherwise it is just facilities management. A manager may indeed be the one to think about backups and grid connections, but not about the changing cultural position of those. An inhabitant just wants to go about their work. An architect might argue how building technology has long been standardized to a science and made culturally invisible. An engineer might object that automated building management systems already respond to changing electric grid conditions well, whether or not anyone cares to call them smart. Yet the cultural imagination has seldom gone beyond saving a bit of energy here and there, and that is not interesting. Even where neighborhood resilience, and not just individual green building credentials, has become an important aspect of a site's cultural image, a building owner has higher priorities about the appearance, organizational needs, and scale of built space.

So to continue laying conceptual background, there must be some larger, more open questions here. Should infrastructure ever be visible? Since a building has network infrastructures of its own, often beyond the reach of a centralized smartgrid, how are those a part of its use and identity?[4] Given how architecture also invests energy into its construction, which then returns that investment in passive advantages, how to get past assumptions that energy is about operational efficiencies?[5] Now with the Internet of Things, how do tangible interfaces make infrastructures more knowable?

The right way to qualify an argument for architecture's grid edge begins with the importance of scale. Scale makes architecture: the very idea of architecture admits that a system might not be arbitrarily scalable, and that making sense of life depends on well-fit dimensions to purposeful places. Recall, from the opening theme on archipelagos, how architecture must emphasize an intensification of design and experience within fixed spatial limits, amid dense clusters of others simultaneously doing so differently. The alternative is all flux. Without particulars of scale, the smart city does not seem so civilized.

That has not remained as obvious as it should. Amid times otherwise too ready to imagine everything through small handheld devices connected to vast distant data clouds, any other intermediate scale or more physical situation receives relatively less consideration.[6] When most systems and networks seek to be scalable, it can be forgotten how one particular scale might have more meaning or fit than another. At least that remains so in physical space. Human embodiment remains the measure of much, and for form and space, the orders of magnitude closest to human scale remain where design does the most cultural good.

That is the aim of this argument. This is why it is crucial to bring the built environment into many more infrastructural conversations. In order to make better use of infrastructures, design is not only the networks but also the perimeters and access points. These have presence. Physical, spatial configurations of objects, props, rooms, building enclosures, streets, and neighborhoods help make sense of what happens where in life. Hence it seems regrettable to reduce life to the individual device and the collective cloud. Many important scales exist in between (figure 5.2). At each successive wrapping on this spectrum of scale of spatial boundaries, design opportunities arise in how networks do and do not cross them.

huge and distant

regional economy, ecology, identity

ownership and control by network platforms

citywide infrastructure

neighborhood, campus, district

social configurations and insitutions

streets and courtyards

building enclosures

floors and building zones

human scale in architectural space

specialized rooms

fixtures, fittings, furniture

obsession with personal apps and devices

physical tools and props

familiar objects

small and present

5.2   The importance of intermediate scale; not all is devices and clouds.

How so too with local electricity? The old assumption that one small meter divides the two quite-separate domains of the grid and the building may not work so well anymore. The way that buildings involve many networks of their own deserves new consideration. As sets of devices, systems, and networks in themselves, buildings have their own, more local realities, whose interoperations are increasingly adaptive. No longer just costly building equipment automation, building networks increasingly adapt according to uses and users, and quickly pay for themselves in the value of data analytics.[7] Not just a middle scale between countless small devices and one single grid, these operations tend to have their own locally institutionalized scales of networks within networks. Since architecture operates socially, politically, and culturally, this is not just a technical conversation.

So while this will take some unpacking, already here are some keywords for the argument ahead: material systems, tangible interface, boundaries and networks, social infrastructure, and intermediate scale.

The Origins of Grid Edge

"Grid edge" (an expression which has approximately a hundred thousand Google search results at this writing) was coined in 2013 by the energy market news analyst Greentech Media. It did start something good. Today the business council, annual conference event, and awards program that Greentech Media began under that name now stand among the better resources collections in a boom. Already a useful category, grid edge comprises all the tangible, local, dynamically adaptable aspects of interface between the public electric utilities and local sites not only of use but also generation, sharing, and occasional independent operations. This replaces the conventional meter, and blurs what is or

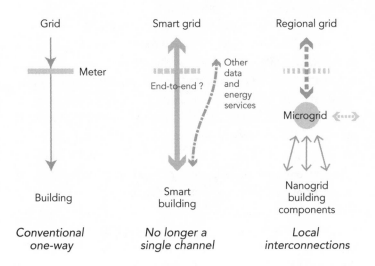

Grid   Smart grid   Regional grid

Meter

End-to-end ?

Other data and energy services

Microgrid

Building

Smart building

Nanogrid building components

Conventional one-way

No longer a single channel

Local interconnections

5.3   Blurring the meter does not mean a single end-to-end smartgrid; buildings and districts may increase their network independence.

is not "beyond the meter." It brings design prominence to physical components with embedded information systems too. Grid edge is local electrical interfaces that you can own, touch, and interoperate.[8]

As interpreted in Greentech Media's oft-cited 3-D infographics (too detailed to show here), grid edge emphasizes the physical assets, network controls, and data applications of local distribution. Increasingly these communicate not only within central utilities but also individual building networks and a general Internet of Things (figure 5.3). Grid edge thus provides a richly diversified interface between utility and "consumer/building/distributed energy resource." A layered model of physical assets, network controls, management apps, and data analytics replaces the single point of connection from building to grid. Grid edge technologies intervene between owner and utility to create a well-managed zone of distribution that supplants old distinctions between "last-mile" utility distribution and individual sites "behind the meter." Such a zone may belong to the utility, a single campus, or a new kind of microgrid institution.

Many grid edge elements belong to particular sites and buildings: physical fixtures such as solar panels, batteries, or charging stations; control devices such as sensors, chargers, or inverters; direct current power networks; responsive building systems such as adaptive facades or programmable lighting; management apps such as energy monitoring, vehicle-to-grid connections, or building operations; and use analytics for performance, utilization rates, and two-way exchanges. Most of those can disappear into building design, as so many other technologies have already done. Yet their presence brings the incalculable cultural value of electric power back into the open. Smart enough is different enough not only to operations but also to sensibilities. While not so great a change as the first elevators, cool lighting, or plug-in appliances, it may not be invisible either. Physical grid edge elements provide tangible, knowable tokens of reform (figure 5.4).

Otherwise the agenda simply becomes smartgrid on a building scale. As the British company named Grid Edge offers, "Our AI software turns public and commercial buildings into intelligent and optimised energy systems that reduce energy costs, cut carbon emissions and transform on-site energy technologies into revenue-earning assets." This extends the role of building automation systems, and those are millions of networks quite distinct from any end-to-end notion of smartgrid. These have already become normative, per Wikipedia: "Almost all multi-story green buildings are designed to accommodate a BAS for the energy, air and water conservation characteristics. Electrical device demand response is a typical function of a BAS."[9] Today this role expands to accommodate two-way energy management, dynamic pricing, carbon taxes, electric vehicles, and so on.

5.4 In the open, at the scale of a small building as a visible token of grid reform: a fuel cell on the University of California at San Diego campus microgrid.

As building electricity assets and costs diversify, smartgrid starts to pay for itself at the scale of the large individual building, often a commercial or institutional building, often for the retrofit of older, more wasteful buildings.

Building operational analytics seldom stir the mind as architecture. So long as these are network control systems run by a few facilities managers and mostly invisible to everyone else, little cultural interest surrounds them. Of course they have benefits, and those can be palpable to everyone. According to the World Green Building Council, health and wellness remain the number one motive for green building, ahead of energy savings, and certainly ahead of aesthetic appeal.[10] Health and wellness presumably begin from better light and air, but they also arise from more freedom to move around, more daily or seasonal flexibility of configuration, and more inhabitants having a way to tune the momentary qualities of the environment.

Somehow there is design opportunity here. No longer just end-use appliances on a centralized utility network, not just about operational efficiency, and definitely not dematerializing into a cloud of digital media, today's increasingly networked built environment deserves newfound appreciation. It deserves cultural speculations and criticisms. It needs to be understood not just through so many clever features and apps but also as an embodiment in everyday habits. Today's built environment deserves newfound appreciation for its balance of passive performance, networked automation, and wondrous activation.

In an oft-cited recent case, "The Edge" happens to be the name of a prominent building in Amsterdam, which at its completion received the highest Building Research Establishment rating yet given (as of 2016) to an office building, resulting in proclamations of "the greenest building in the world."[11] Deloitte, its main occupant, which as a management consultancy is skilled at estimating return on investment, explains that while the building's technological features will pay for themselves in under ten years, nevertheless the greatest value of the space is in workforce productivity, as based on comfortable social participation, and choices rather than the invisibility of surroundings can be a factor in that. According to its architects, the design "starting point was the design of a social condenser."[12] As is increasingly popular, this set up an open office system with strong visual and atmospheric linkages among its set of available desks, meeting spaces, and breakout sites. For this it provides personalized environmental management, with local light, heat, and blinds all adjustable by smartphone app, and personal preferences maintained as people move about (figure 5.5). "A survey found that while fewer than a quarter of employees actively use the app's thermostat features, three-quarters say they love it."[13] Apparently the ability to

5.5 Interactivity in power-over-Ethernet lighting. Courtesy of Philips/ Signify, 2018.

move around and tweak has increased the satisfaction with the overall conditions more than it has led to squabbles over settings. Lighting is the most palpable edge to this system: every fixture is individually addressable by the system, half of them have sensors, and LED power needs are low enough to be delivered over an Ethernet. Heating and cooling draws from underground thermal storage aquifers. Solar power generation has been added to the adjoining buildings too, which all share it. Overall electricity savings are as much as 70 percent relative to a conventional building.[14] Although the building name evokes a more general leading edge in organizational advantages, and not grid edge per se, and although electricity itself does not figure in reports about the space, a heightened, participatory awareness of building atmosphere and performance certainly does.

Hybrid Components

In light of these many agendas, consider architecture's role in three ways: first through digital-plus-physical hybrid components, second as a topology of access, and third as a frame of social practices. Grid edge normally implies distinct components and not whole spaces. Some of those components are hidden away in control rooms, and often the most knowable component is one's own smartphone, now a control device for space too, yet other components are increasingly built into objects and contexts of use. Despite recent obsessions over mobile apps, there are situated technologies too. Some of those may appear in the new delights of tangible interfaces. When presence or gesture is input enough, that can have effortless appeal. Of course, it must also have use, sometimes in the context of physical activity. For instance, digital-plus-physical hybrid bike shares and scooters demonstrate as much at street level. This is a fundamental principle of interaction design: information technology disappears into everyday things that are not primarily thought of as digital. At a door, for example, you might tap an electronic key fob for entry. On a factory floor, a machine might store process histories, operator preferences, and current job queues. In a hotel lobby, interactive fixtures and furnishings provide an array of amenities. Physical-plus-digital hybrids pervade the built world from the most mundane motion-sensing streetlights to perennial attempts at smart homes to public interactive media facades to actuated green building skins.

Consider, for instance, the appeal of programmable glass (figure 5.6). At the moment this material system is found in the roofs of some cars and a few select office settings. If its economies of scale advance at all like photovoltaics have recently done, the effects could be palpable. Using an electrochromatic process, which occurs in nanoscale

5.6 Actuated green building skin: programmable electrochromatic glass. Courtesy of View.

overlays that have been applied to conventional plate glass, it is possible to alter the opacity of the surface on demand. Using physical actuators driven by wall-mounted controls, it is possible to embed such control into a room, much as is already normal (but would itself have seemed astonishing little more than a century ago) for lighting and blinds. Then using handheld apps and building networks, it is possible to introduce predictive control and even adaptive learning. Besides being quickly changeable on demand, the opacity of a facade can slowly vary with the weather, the days, and the seasons. This more ambient quality can afford efficiency, wellness, and presence all at once. Unlike blinds, the tinted glass maintains views. If ever such technology becomes cost competitive with conventional sunshades and blinds, it would quickly grow usual enough to afford a variety of subcultures and practices. If this sounds astonishing, so quite recently did the notion that solar power could become cost competitive.

Since designers are in the business of proposing what does not yet exist and imagining how else things might be, digital-plus-physical prototypes sometimes do go viral: speculations cast aside conventions, early adopters suspend disbelief, clichés perpetuate, and soon it seems like *Star Trek* science fiction. Or perhaps it results in a chirpy device in your kitchen.

Instead, some designers start from *social* fiction. What if there were cash machines at the supermarket and not just the bank? That was not a technically indicated solution. What if it becomes acceptable or even fashionable to go about wearing earbuds, the promotors of the iPod once famously speculated? What if office space becomes a flexible, data-rich platform, as WeWork first ventured. Or, in the social provocation of the "Great Transition," what would it be like to have more local energy independence than has been known for hundreds of years, and live in synch with the days and the seasons?

The social dynamics of adopting a technology seldom move in a single direction. Cultural adaptations can both drive or be driven by new technological possibilities. How they make change knowable depends on a wide flexibility of interpretations and only later closes out such social experimentation. These are core principles of science, technology, and society narratives. It is worth repeating that the circumstances of electrification have long been a core theme in such studies, and the time has come to attempt such narratives on clean smart green micro energy.

As ever, new on-site artifacts serve as valuable tokens in that kind of work. For example, earlier technoutopians might not have anticipated the appeal of camouflaged rooftop solar panels—a product prominently brought to market by Tesla, today's most prominent electricity design futures company (figure 5.7). Although explained as if the seamless integration of roof and solar panels is just one more sense of visual appeal, the camouflage presumably also serves a social purpose at relieving peer pressure (whether to outgreen the neighbors or hide green among deniers), conforming to

5.7 Tesla roof tiles (2016), a grid edge building component. Courtesy of Tesla.

neighborhood appearance guidelines, or enacting some new wonder at tangible interactivity or simple activation itself.

Except with all due wonder at those solar roof tiles, by most accounts the more unpredictable necessary change is in storage. Batteries are now the most vital technology, the killer app, and the key to the grid edge. Like information technology, batteries get built into things not primarily thought of as batteries. Things that run on stored electricity can also add storage capacity to their network when not in use. That works best on a manageably smaller network. Rooftop solar alone does not do much good in an outage. To run as an island, a site needs storage, alternating and direct current conversions, and power management too: a microgrid.

In that pivotal year of 2016, when Tesla, the vehicle builder, merged with SolarCity, the photovoltaic panel maker, Elon Musk explained the importance of making microgrids more accessible. Individual owners who lack the time, tools, or skills to do the kinds of analyses that larger institutions use in planning local electricity need a simpler "click-and-buy" microgrid product.[15] While that may lack the diversified generation or demand response that give other microgrids such advantage, it could provide enough technical and social convenience to let home solar plus storage plus electric vehicle soon scale up. Futurism aside, part of the immediate appeal here is in the architectural components. It is not an accident that the battery storage component so pivotal to this scheme is named Powerwall. This is an example of using an architectural move to trigger a shift in cultural expectations. While a virtuoso piece of engineering, it is not just a technological solution.

The design academy easily imagines more. For instance, many academic architectural researchers have prototyped sensate, kinetic material systems.[16] Elsewhere, conferences on urban media facades have shown the integration of display with enclosure and the importance of interactive public art. Any number of conferences on smart green building now study responsive building components. Yet too often the focus tends either toward design novelties or, at the opposite extreme, performance metrics, and not often enough on the middle ground of everyday cultural experience. They often imagine new infrastructures, but think less about making infrastructures knowable.

Conversely, the engineering and policy groups advancing the meme of grid edge have not got much focus on the experience of space and place. Their use of the word "architecture" does not emphasize buildings. It would be difficult to delve very far into the sociotechnical dynamic of bridging these worlds of engineering and policy with those of architectural proposition and everyday interactivity. It would be easy to dismiss the gap as more smart green blues. The point here is that new hybrid components, such as the much-hyped mid-2010s' Tesla products, now invite a deeper science, technology, and society interpretation, and the wonder of responsiveness can be the best way into that.

Topologies of Access

For a second role of architecture by which to rethink grid edge, consider the reciprocal relation of boundaries and networks.[17] No matter how networked life has become, not all is flux; fixed architectural circumstance still matters. Thus to understand the possible pertinence of architecture to the microgrid boom, it generally helps to think a bit more about the importance of access points. A boundary gets more interesting where it lets some networks across. A network gets more interesting as it crosses boundaries. The point of connection becomes a vital object of design. Good design not only filters but also dignifies boundary crossings. Fixtures and fittings may not be an infrastructure's most complex, costly, or controversial elements, but their experiential qualities give it a face.

Recall some access points not made obsolete by the internet: doorways, loading docks, water mains, gas meters, security cameras, Wi-Fi routers, package drop boxes, food courts, teleconference tables, privacy curtains, bug screens, light gratings, storage rooms, meeting points, guard desks, signposts, and campus gates. The word "fixture" indicates a component device attached as an accessory to a system. A "fitting" is an attachment that has been properly and felicitously adjusted to its circumstances. Thus a tradesperson who assembled custom local connections to a steam network was called a "fitter." Using such language, it is only a bit of a stretch to say that architecture "fits" an organization and its set of uses to a site, a microclimate, a social milieu, and a set of network services. Yet it would seem odd to describe a building as a "fixture," since it is a complex, multitudinous affair and not a widget. It might seem odd even to just describe a building as a *set* of fixtures, as if, say, stairwells and roof gardens were fixtures. Instead, some emergent spatial character, enabled by so many networks, boundaries, and fixtures, is normally the object of design. So it may be counterproductive that architecture's electrification has been reduced to such standard expectations conceived as a set of equipment. Conversely, the higher aims of architectural character have seldom (although more so lately) involved fresh expressions of network infrastructure.

"Walls, fences, and skins divide; paths, pipes, and wires connect," architectural futurist William Mitchell wrote as the internet ascended at the turn of the millennium. Not only the idea and structure of a network but also its important reciprocal relation with boundaries repeats across a spectrum of scales, from body to room to building to street (or campus) to city to region to a worldwide network of networks. The more these scales interconnect, the more any one becomes a design opportunity. As Mitchell declared, "Today the network, rather than the enclosure, is emerging as the desired and contested object: the dual now dominates."[18] But this is not an entirely virtual arrangement, nor one experienced entirely though smartphones, because so many connection points are built into physical space—indeed more are embedded there all the time. Nor is it all one uniform space (isotropic or panoptic, to throw in some usual jargon) all controlled by a faraway cloud. Some hierarchy, topology, and perhaps even privacy remains. Physical scale and circumstance remain not just as a legacy but also for the increasingly vital reasons of managing, disambiguating, and occasionally taking respite from the escalating flux of information.

The point about boundaries and networks is likewise twofold: people increasingly identify with networks, but the scale at which they do that, or more specifically the embodied circumstances from which they do that, remain essential to those practices. Of transportation, Mitchell wrote, "By the mid-twentieth century, though, the most memorable ideogram of London was its underground network, and that of Los Angeles was its freeway map; riding the networks, not dwelling within walls, was what made you a Londoner or an Angeleno."[19] Of social media, he emphasized "placing words" in physical situations: "Literary theorists sometimes speak of text as if it were disembodied, but of course it isn't; it always shows up attached to particular physical objects, in particular spatial contexts, and those contexts—like the contexts of speech—furnish essential components of the meaning."[20]

Tapping into the city electric grid was once one of those definitive network experiences, which citizens identified with by means of the physical situation of use, especially back when only cities had electricity. Across a scale from device to room to street, electrical activation became an emblem of modernity. The expression "grid edge" did not exist then, but as investigated earlier, the use of a streetcar, radio, factory power tool, or domestic light switch was part of a new experience of being modern—one that had yet to fade into invisible, universal normalcy, and one that came with social cues and connotations.

Today instead the experience on offer is smart and green. Not just about efficient operations but also about the identity, mythology, or brand of its owners, the smart green building stands for something cultural. That may be about a culture whose highest aspirations are expressed as engineering solutions, or it may be about a culture that is quickly adapting toward more participatory resilience. The individual health and wellness benefits of greener building may be only just the start of larger cultural change.

Frames of Social Practice

Architecture provides cultural context, especially through institutions, and does so with new technologies layering onto, rather than replacing, many earlier ones.[21] As has been argued at length in recent years, partly as a defense against anytime/anyplace abuses of smartphones, one way that architecture gives cultural context is through the disambiguation of messages. For instance, social speech tends to rely on contextual cues; one would not want to speak out of place. Moreover, many specialized or especially institutionalized spaces provide props and frames not only for speech but also the sending of messages as well as the processing of data streams. To interpret this role, it is common to appropriate the film principle of mise-en-scène.[22] No mere backdrop, the scene configures props, interpersonal distances, spatial scale, and perhaps institutional signifiers to establish how a situation might best play. It is not just in theater that experts play situations rather than merely following rules or needing apps for everything. It is not just in crafts, sports, or dance that a skillful practice, often one valued in itself for the sense of flow it provides, makes active, nameless use of context. This is commonly called *the extended mind*.[23]

Unsurprisingly, many more designers thus take interest in the cognitive appeal of activity amid palpable things, and especially on the latter, design interests have surged on the phenomenology, new ontology, sensate autonomy, and even imagined subjectivity of things.[24] So the wonder at activation is alive and well here. Architecture's connection from smartgrid to everyday practices occurs through palpable engagements of situated things.

Yet this is not all individual handheld smartphone apps. For all today's emphasis on individual choice and mastery, nevertheless social practice gains greater benefit from well-configured surroundings and is often the basis for building projects. Architecture represents organizations to their constituencies. It reflects shared aspirations, legacies, and customs. More so than any app, and least more subtly and persistently so in the long run, often through the habitual, internalizing effects of activity, the built environment shapes sensibilities. Life has not disappeared through the looking glass into what once was called cyberspace. Indeed the rent is higher than ever, and the struggle for competitive advantage, quality of life, and fashionable expression of organizations to constituencies increasingly depends on smarter, greener space.

It would be a stretch to compare the social will to be modern, back then, with the will to be green today. It would be an oversimplification to say that electrical appliances then and smart power apps today have powerful social constructs behind them. Nonetheless, a new social history of smartgrid might admit that the connection points to infrastructures gain cultural resonance from their built environmental circumstance. Architecture's grid edge becomes more than a set of devices. Cultural change comes through shifts in embodied practices more so than from instruction by apps or new technology on the roof. Change away from command-and-control uniformity seems welcome in this regard (and many others). Locales may differ, and second only to the diversity of persons themselves, that experiential diversity, so compromised by late modern standardized building, now opens out once again—perhaps even for the components, boundaries, and social frames of electric power.

"The fact that relations between infrastructures and practices are multiply mediated means that there is (consequently) scope for substitution, adaptation, and dislocation," sociologist Elizabeth Shove has observed, adding,

The "same" practices can be enacted in much the same way despite sometimes significant differences in how resources and services are provided. As a result, changes in infrastructural arrangements (e.g., from centralized to decentralized power supply, or from local to more distant forms of water treatment etc., are unlikely to show up *in* practice unless they affect the "proper" functioning of one or more mediating devices. By the same token, individual practices and mediating devices may well change, sometimes radically so, but with little or no tangible impact on the infrastructure on which they multiply and variously depend. Among other things, this means that analyses of the relation between infrastructure and practices need to consider the *mediating* function of appliances—broadly defined.[25]

For an instance in the emerging social history of local electricity, here return to another aspect of the largest smart green neighborhood to date—the former Mueller airport site in Austin. Here, beside the smartgrid proof of concept in domestic use, there are several prizewinning commercial buildings. Most apparent among these, the H-E-B supermarket has gained wide use and admiration.[26] Designed by the distinguished Texas architecture firm Lake/ Flato in collaboration with the soft energy think tank Rocky Mountain Institute, this project was completed in 2013, and received a national top ten environmental award from the American Institute of Architects in 2016. Although conceived independently of the Pecan Street project, this market has become an important anchor and driver of development on the Mueller site, and also an especially tangible everyday encounter with new electricity practices for residents there. It makes at least some of that more experiential. For example, instead of the more usual shock of passage from excess heat to excess air-conditioning, this site provides a transitional set of microclimate zones—one amusingly known as its "Texas-sized vestibule." Given that a food market is among the most electricity-intensive commercial building types, and refrigeration is the bulk of that demand, the project features innovations on refrigeration. For the tangible access point to this system, the design applies a new format of display case instead of the chilled aisles and open-ice, topless freezer bins, thus saving on water too. Although not integrally linked into the Pecan Street microgrid prototype on the Mueller site, this project likewise tracks changing habits of use, performance, participation, and appreciation of daily and seasonal cycles.[27]

For a different scale and use, it is worth a look at one more case of grid edge in a project known for responsive building components, without regard to local electricity operations. The New York Times Tower, completed nearly fifteen years ago, still illustrates how best practices in smart green building enable new practices and sensibilities.[28] Here was one of the most prominent early experiments in responsive building skin (figure 5.8) in a major project by a signature architect Renzo Piano, who is known for elegant sunshading devices. To test the new system, the project built a full-scale 4500-square-foot outdoor mock-up of a corner bay, with over 100 sensors reporting in real time to analysis and simulation software at the lab. The technology combined exterior adjustable sunshade louvers, mechanical interior blinds, and individually programmable lighting fixtures to create an adaptable system that automatically responds to changing sky conditions, and can be reprogrammed whenever interior uses and floor zones change. In the visible result, *Metropolis* magazine observed that "the design for the 51-story tower, featuring an 800-foot shimmering glass curtain wall, promises to bathe notoriously cranky reporters and editors in natural light."[29] In evaluations conducted after five years of use, three-quarters of those surveyed were happy with the lighting, and close to half were more than neutral about the overall system (which is a higher result than a baseline conventional building gets), and the need to manually override the system was infrequent, or less than 2 percent of the days. Meanwhile, lighting electricity savings were over 50 percent compared to a baseline nonresponsive standard code-compliant building. Forty percent of the building electricity came from cogeneration with heating,

5.8 The responsive skin of the New York Times Tower designed by Renzo Piano (2004). Courtesy of Renzo Piano Building Workshop.

and the overall heating and cooling loads were reduced by an eighth by the responsive skin.[30] By now such numbers may seem usual enough and tedious to consider, but in the early 2000s, this was noteworthy proof of concept, both of the prototyping process and performance in signature design.

Meanwhile, as Piano's collaborators at the Lawrence Berkeley National Laboratory have explained, the result demonstrates the value of not only sensate physical prototyping but also life cycle design for embedded systems.[31] The project's layering shows a gradient in the physical-digital hybrid elements. The exterior, which must last longest, under the harshest conditions, has the most robust composition, whose electromechanical elements are least likely to need updates, and then software updates only. The interior mechanisms allow but may seldom need new modules. The lighting system, which has the fewest mechanisms and most software, affords the most frequent updates. This alleviates the one usual concern about sensate building elements: physical components with a life cycle of decades must not depend on electromechanical or software components with a life cycle of just a few years. Someone might observe the same for the local energy network that powers them. If light still does the most to make electricity knowable, this responsive lighting system, especially if considered together with the building cogeneration system, was one of the first to make architecture's grid edge culturally recognizable.

Intermediate Scale Reconsidered: Architecture on the Microgrid

"Architecture's grid edge" is not a common figure of speech, nor does its speculative usage here anticipate it becoming one. It is just a potentially useful way of describing a new perspective on local electricity. There is much about architecture that makes grids tangible. There is also much about architecture that is beyond the reach of any centralized smartgrid system. Now that the power grid's economies of scale have reversed, all this seems worth exploring anew. Now that a more variable pursuit of comfort overturns older quests for uniform command and control, much of this gladly regains an awareness of surroundings. Now that the need for resilience obviously overturns older values of efficiency, much of this occurs at local scale. In the process, architecture blurs an old boundary between building and grid—one that was formerly quite clearly marked by the electric meter. Today the boundaries and networks permeate many more components and systems. As a result, architecture's grid edge deserves creativity in itself. Where for a century, the more energy that buildings used, alas the more they all seemed the same, today local difference can be beautiful again. Local patterns of use, physical systems of integration, and material cultures of comforts all provide design opportunity and challenge. Much like a century ago when electricity was new, architecture again seeks to express a new grid awareness and new arrangements of local energy networking. Although engineering, policy, and economics remain more essential to any infrastructure building, the discipline of architecture cannot remain a neutral bystander here. A better building might be a good place to be if the grid goes down. That might be enough of a start. Yet a really good building is not just a place to turn on the lights.

6   SITUATED INTERACTIONS

For most of the history of architecture, technology, and the city, most of the world was static and mute. But then it powered up, and now it boots up too, and under the right conditions it responds. Alongside these advances in activation, habits have changed as well. Today many more situations come with expectations for interactivity. In this, despite mobile devices anywhere and everywhere, context remains a consideration. Dematerialization seems less of a concern than it was imagined twenty years ago in the days of "cyberspace." While handheld media now dominate experience, its location-based apps can be the more interesting ones, and these meet with situated technologies too. Almost anywhere, these changing contextual practices deserve some careful cultural tuning. Back when mobiles first exploded onto the scene, this bit of anthropological wisdom became known as *situated interactions*.

Today the Internet of Things has made many everyday resources somehow participatory. Streams of entertainment, merchandise, food, and mobility all afford personalization, customization, and pattern recognition. So too for the pursuit of indoor comfort. So too for generating, distributing, and using electricity. Today's new priorities for decarbonization and resilience alter century-old assumptions about the electricity grid staying invisible. Although still mostly understood as an always-on resource from far away, and although almost any utility provides convenient apps for monitoring it, even this resource now invites a more local, participatory perspective.

So note an important paradox here: supposed automation often actually requires participation (figure 6.1). Much as in the industrial age, supposedly laborsaving devices instead caused more frequent activities, likewise under the Internet of Things, supposedly smarter things, buildings, grids and cities require much more careful watch. The more that the built world gets infused with digital devices, the more those require monitoring, tweaking, and tuning. The better tuned a local system becomes, the more variations it can manage, and hence the more customization may occur. The resulting experience is indeed situated.

In a world of systems, appropriateness takes calibration. Wherever the responsiveness of a system depends on its circumstances amid other systems, it is difficult to formulate predictably in advance. Instead, experience in its use shows how best to configure, tweak, and monitor it. Often this

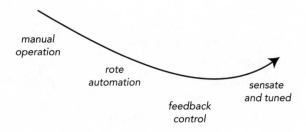

6.1 Paradoxically, increased mediation involves more participation.

process is qualified or conditioned by something else. The more that technologies accumulate, the more they must be adjusted and refined in relation to one another. Time shifting, prioritizing, storing, and backing up become more usual. What used to be set-and-forget now becomes watch-and-tweak. Energy can be felt to matter and not just go ignored. A world that is always on starts to feel more like a world that is at least appropriately on.

This circumstantial aspect of design seems quite familiar in architecture and urbanism, and is becoming so in interaction design. This argument points toward engagements with local energy needing plenty of local adaptation. The appeal of the microgrid meme comes in part from this prospect of participatory tuning.

Yet there are plenty of misconceptions too. Hardly anyone wants one more thing to think about; any new grid awareness must occur habitually in the background, like the built environment itself. Not everything is in the home; there are also shared spaces, whether commercial, institutional, workplace, or recreational, and those tend to have greater capacity for technological innovation, but also greater difficulties in managing participation, but also greater influence on social change. Not everything is technological novelty; as hype cycles show, that soon wears off. Not everything is instruction; most practices occur more namelessly and change more socially on the basis of their material settings. Not all smart buildings use less energy than inert ones. Not all technologies work just as well anywhere; most must be adapted to circumstance.

Thus any work on participatory local electricity must explore changing expectations for interactivity. Between invisible everyday background automation and foreground apps that seldom hold interest long, there arises a middle ground of background awareness and occasional adaptation.

For instance, within the context of on-site solar generation, note this is also a wish for understandable operations. While new expressions of living with local energy might take some time to materialize, already expectations shift toward noticing them. Throughout the production, upkeep, and responsible use of increasingly local electricity, the information technologies that make it go can also make it more enjoyable.

Here are the steps of an argument to follow. Let them lead toward a speculative stance, let that emphasize tuned response, and let that help interpret the microgrid meme. To repeat, this is not an argument that every organization wants to manage its power, nor that microgrids are yet practical or preferable, but only that cultural expectations have shifted toward interactive tuning. Quickly note the general path and then consider each step briefly in turn.

The *qualities of response* have improved enough technologically to become a major cultural pursuit in themselves. This is a salient trait of these times. For this, embodiment in the *Internet of Things* raises the cultural stakes. It brings the physical context back into consideration. More tangibly embedded technologies tend to couple better with deeper *social practice change*; abstract technologies imposed from outside too often have less capacity for that. Today much such change occurs not only amid but also necessarily sometimes against a prevailing *culture of convenience*. Having anything anywhere anytime has proven quite costly. Instead, wherever they can, people increasingly choose occasional, artisanal, and local. Again, having more locally responsive systems takes more *monitoring*, *tracing*, and *tuning*. Some voluntary degree of monitoring and control sometimes provides more comfort than purely passive automation. This is worth repeating: much as industrial laborsaving devices paradoxically caused more chores, so today in an interactive era, responsive environments paradoxically take more consideration. So far for local electricity, that control is mostly manifest in *dashboards*: graphical interfaces for owners of local electricity systems, which play a complementary role to big, distant control rooms in power plants. Yet more about that presence could be more tangibly built into the form and appearance of architecture's grid edge. Habits of engagement with physically present manifestations can eventually give energy presence, so that is less likely to be ignored, abused, or wasted as if a free and infinite good. *Interacting with electricity*, so to speak, thus seems poised to become a distinctive kind of cultural expression.

Qualities of Response

Responsiveness has become a cultural aim in itself. Cultural expectations for perpetual interactivity surely must count among the most salient traits of the times, by which these times most differ from any previous ones. People want responsiveness for its own sake; look at how the smartphones come out at the slightest pause in experience. Obsession with strokable devices goes beyond all their usefulness, however unprecedented that usefulness may be. Almost any urban scene has someone stroking a screen or waving a device at something. If an activity does not yet have an app for it, either imagine that a company is out there building that app or else, if there are reasons why that is not possible, imagine that activity fading from everyday life.[1] To interpret this sea change in human subjectivity may take the best philosophers. This is now the default sensibility.[2]

Responsive objects, spaces, and systems do permeate everyday life. Walk into a room, and the lights come on. Sit back with a trackpad, stroking it for the immediacy, without much concern what about, since it all gets responsively filtered to your tastes anyway. Open a suite of apps on a shared workstation, and it loads your preferred personal settings. Tell your countertop device that the kitchen needs some more towels, and it orders them. Glance at a product display in a window, and its face recognition software adds that glance to its exposure count. Step up to a piece of public interactive art, and its output patterns may shift, as its adds you to its crowdsourced data inputs.

These new sensibilities show up at street level, for instance, where many more people discover the joys of navigating the city. That perhaps begins from the instantaneous advantage of maps delivered to the very locations they describe. Then it starts layering and filtering, say, for traffic density, nearby friends' locations, or previous itinerary histories. Next it starts annotating: tagging places with opinions and pointing to new places based on recent activity. As a result, navigation becomes a more open process, not necessarily about destinations. Anyone who has grown up with the responsiveness of gaming might find a little play in the city as well. Agency is enjoyable.

Of course not everything responsive is electronic. Naturally the most satisfying engagement comes from skillful practice with simple favorite implements, such as a sharp kitchen knife, a good pair of skis, or a fine violin. Yet the qualities of transparency and flow found in the likes of cooking, sports, and music sometimes also appear in the use of systems. For instance, a fine driver's car is said to be responsive: its balance, elasticity, and dynamics matter more than its touch screen.

The response of a system enacts many first principles of engineering, and many of those in time. How quickly, directly, and reliably does a system reach a desired new state when conditions or settings change? When does a perceptible transition feels better than an instant step change? Where do elasticity and rigidity play complementary roles? Well-made handles, buttons, and dials demonstrate these design factors, for instance. Even in software, better design transitions in sliders, zooms, and overlays can bring a richer sense of fluidity to operations. Of course in digital media (far more so than driving a car, say), that fluency creates an unprecedented density of possibilities and variability of control. Configuration becomes a pleasure in itself. All this has made interactivity into a preeminent cultural category.

Almost by definition, a system organizes and operates a set of objects to achieve a controllable response. Normally this is understood as a process: a system has intent, protocols, and controller inputs. Yet it must also be understood as a material instance: a system is one of many systems, in the world, embedded into things.[3] Moreover, by principles just explored for the case of architecture's grid edge, a material system may perform in ways just as dependent on its geometric configuration as on any properties of its components or the parameters of its algorithms. As is quite important to the overall argument here, everyday engagement with responsive material systems also can develop strong cultural patterns. To gain any empathy, it must be embodied in practice.

Internet of Things

Just keep repeating: not all interactivity is on handheld devices and distant server clouds. At least as much information technology gets built into sites as gets carried about in pockets and bags. Much of it operates those surroundings. Industrial control systems have been around for a century for example, and were the origins of cybernetics. Thus as briefly explored in the histories that opened this project, smart predates digital, and of course information technology became pervasive long before the arrival of smartphones.

By most estimates, microchips have outnumbered humans on earth for as much as a quarter century. Increasingly they connect: not everything is or needs to be on the Internet of Things, yet ever more things seem to be so.

Sensing, processing, storing, and communicating have by now been built into all stages of the electric power infrastructure, and especially into the rapidly increasing range of devices at grid edge. As just noted, the utility meter that was once the clear edge has become the most immediate example of this; today's building networks and microgrids blur that boundary, and can become important, variable, grid edge components in themselves.

In pursuit of rewarding engagement, then, not all is tapping on screens. Embodiment in objects allows a more intuitive grasp, more effortless attention, and richer response (figure 6.2). Thus the discipline of *tangible and embedded interaction* has advanced rapidly in recent years.[4] A design research conference by that name began in the aughts and now attracts hundreds of participants each year.

6.2 Embodied interactivity, here in a touchable lighting tile system, occurs without names, procedures, or menus. Courtesy of Sensacell.

In a fundamental axiom of good interaction design, more rewarding response engages more of the senses while imposing less of a cognitive load. Many public art installations do this well. For example, *Ascent*, a sidewalk piece by Jen Lewin, cultivates an awareness of the daily cycle of the sun. There are no options to click. Running on solar-plus-storage itself, it processes and reflects surrounding colors by day, and glows around footsteps by night (figure 6.3).

Such delights of nameless engagement remind that not all is apps. Good interaction design can mean something else than attaching names, features, and instructions to everything. Instead, it develops more nameless knowledge. Often that involves body language, haptic orientation, masterful skills, communities of practice, and the scale and configuration of their props and contexts. Cognition researchers sometimes

6.3 Awareness of the sun, in Ascent, an interactive installation by Jen Lewin. Creative Commons.

call this the extended mind, but practitioners often just call it craft.

As the cognitive science of engaged activity has regularly shown, physical props, spatial configurations, and embodied scale all contribute to knowing, often in a nameless, app-less, socially situated way that flows. Because bodies condition knowledge, and interpersonal presence conditions practices, this is important to investigate from the perspectives of anthropology, phenomenology, and sociology, and not only as engineering.[5]

This is the upside of the Internet of Things: use of physical form and context makes activity more intuitive. Whether in wearable sport media, urban wayfinding, sensate architecture, or public interactive art, the prospect of tangible response to physical movement provides an enjoyable alternative to so much obsession with screens. Embodiment helps activity flow without procedures and names, and that makes it more effortless, and that in turn makes embedded information into one more component of a tangibly present situation.

If the goal is to understand and make better use of otherwise-invisible infrastructure, then its fixtures, fittings, and edges need presence in active practices of everyday life. For example, habits of sorting household recycling cultivate awareness of the waste stream infrastructures out there. By contrast, electricity is too cheap and too abstract to afford such little rituals.[6] Moreover, energy savings has been cast as a rational, mental approach rather than an embodied habit. As ethnographers Grégoire Wallenborn and Harold Wilhite have cautioned, "In mainstream theorizing about energy consumption, body is collapsed into mind and the demand for goods is both disembodied and decontextualized from social and material worlds. Yet, bodies are repositories of a unique and explicit form for knowledge about the world and this knowledge affects the ways we consume."[7]

Thus the more that the Internet of Things accumulates at an inhabitable scale, in places where it is difficult to opt out, the more it requires careful, participatory tuning. Especially at those intermediate, inhabitable scales between small device and huge network, where embodied in rooms, buildings, districts, and neighborhoods, increasingly responsive spaces reward some new sense of participation. Again, the more information technology gets embedded into surroundings, the more variations become viable anyway. If technology is not to seem out of control, having a least the occasional opportunity to monitor, adjust, or opt out becomes all the more important.

After all, there might be limits to what you want machines to do on your behalf. You are probably content, say, to have passive devices watch for smoke while you sleep. At work, you might have algorithms filter your mail. In a more active process, you might gladly delegate tasks to mechanical devices, such as timers while cooking, rough cuts in woodworking, or an automatic transmission in your car. You may still be undecided about letting a fully automatic vehicle drive for you, however, and it would be another thing altogether to send such devices on your behalf instead of going someplace yourself. Likewise, for instance, in social media, presumably you would not allow an emotive algorithm to decide on the wording, frequency, or recipients of personal text messages from "you." These are cultural and not merely technological concerns.

Culture of Convenience

Among shifting expectations, few have such strength today as the desire for convenience. The wish to have anything anywhere anytime has become distinctive enough in itself. This is costly, but that is another argument. Here the point is

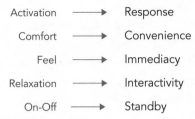

Activation	⟶	Response
Comfort	⟶	Convenience
Feel	⟶	Immediacy
Relaxation	⟶	Interactivity
On-Off	⟶	Standby

6.4   Some conflated expectations in the culture of convenience.

that having ever more about life conveniently on standby is considered comforting (figure 6.4). Like electrification before it, interactivity has come to seem more natural than its absence. To have personal options in the degree of participation or automation thus seems especially characteristic of the times.

Consider some perhaps-familiar descriptive memes often used by sociologists. "Culture of convenience" emphasizes a preference not so much to save time on everyday ongoing tasks as to not lose time to additional isolated tasks. This culture shuns unnecessary, unanticipated actions, unexpected events, or being bound to a plan, especially someone else's plan. Food has been the most usual topic of studies; convenience foods do not save time so much as they save planning ahead. Many such fundamental resources now operate in the moment. Through invisible, complex but timely logistics, and in response to the simplest possible apps, "on-demand services" provide "fulfillment" for almost conceivable task.

Much as Alvin Toffler's famous sociological theory of future shock first described a new degree of inability to keep pace with change, so Doug Rushkoff's theory of "present shock" names the inability to keep pace with the always-on demands of the moment. This ultimately implies an inability to remain aware of any past or future at all, unless assisted by external apps.[8] Hence, in a related meme, the "quantified self" emphasizes dependence on measurement and analytics for knowing (and life hacking) how to live better, perhaps with less inconvenience, and often with more productive personal scheduling.[9] A younger "app generation" understands ever more about life through such technological mediation.[10] This is not so much a lament on distraction as simply an admission of new epistemology. As ever, the world becomes knowable mainly through what you can do in, with, or for it. What has changed is the quality and locus of the response.

In a long-term cultural history of the idea of convenience, written long before the rise of smartphones, sociologist Thomas Tierney observed how convenience was not originally such a personal, consumerist, or time-saving notion. The word came from the Latin *convenire*, which meant "to come together, meet, unite, agree, fit suit." This was less about availability than appropriateness. As Tierney put it, "Something could be described as convenient or as a convenience if it was in accordance or agreement with something such as nature or 'the facts,' or if it was suitable or appropriate to a given situation or circumstance, or if it was morally appropriate."[11] But then modern technology created a culture of transcending the limitations of the body, particularly the limitations of space. As urbanists ever since have observed, technology was said to annihilate distance, and design was said to shift from the configuration of space to the configuration of time.

The era of early electrification synchronized everyone's time, whether to catch a commuter train, speed up an assembly line, or listen to a radio program. Lewis Mumford famously opened his master work *Technics and Civilization* with the claim that the clock, not the steam engine, was the key machine of modernity.[12] The more that the city became organized in time, the more time was to be saved by countless new forms of efficiency, often the efficiency of movements, in which the organization set the pace of the individual. Convenience became a thing to buy, frequently a laborsaving electrical device to own.

Expectation for convenience, of a kind, soon breeds more such convenience. It also tilts cultural dispositions toward that and not other expectations. The way that the smartphones come out at the first momentary lull shows significant cultural change in what constitutes convenience. This is the stuff of everyday network sociology. Public intellectuals such as Tim Wu, Ian Bogost, or Sherry Turkle, to name a few, often lament this condition, and sometimes ask how it shifts other cultural norms.[13]

"There is no way back from the pursuit of convenience," Elizabeth Shove has explained. For today convenience has become about personally time shifting as much as possible to one's own schedule, disposition, and productivity cycles. This may at least raise awareness of time shifting in other aspects of life—for one, a grid awareness about daily cycles in electricity supply and demand—but conversely, it reduces tolerance of delay or intermittency. In the personal practices that Shove has studied, "the spiral of convenience" escalates the time shifting: the more tasks that have become asynchronous and personally managed, the more others now need to become so as well.[14]

Note how the same kind of time shifting that perpetual personal interactivity emphasizes is also what smartgrid emphasizes, as it time shifts various kinds of supplies and demands. Whereas the grid was always instantaneous, now the rise of storage for solar power introduces time shifting even here. In this regard, a disposition toward lifestyle data monitoring and analytics invites a new grid awareness too. The quantified self movement also quantifies resource flows. Writing about smartgrid adaptation, after dozens of case study interviews, Jennie Stephens, Elizabeth Wilson, and Tarla Rai Peterson observed that "the explosion of smartphone usage and expectations of constant connectivity have revolutionized cultural expectations regarding data, information, communication access, and availability."[15] Here, with respect to social practices, and in the context of situated interactions, you might add some expectations for more participatory tuning.

What has all this got to do with the microgrid boom? Don't forget the most obvious relation between the culture of convenience and infrastructures of electricity. Nothing has become more annoying than inconvenience, and there is simply nothing more inconvenient than having the power go out.

Monitoring, Tweaking, and Tuning

The microgrid boom depends on situated interactions. It introduces an intermediate scale between the device and the cloud, and likewise between the nearest light switch and the distant thermal power plant. It takes less for granted and seeks more participation in everyday resource flows. It depends on a local diversity of digital devices. Its sensors, data analytics, on-site equipment, and operational dashboards make electricity more manifest than a more distant standard utility service ever was. Through physical presence in built space, these situated technologies invite better habits of monitoring, tracing, and tuning.

Of course not all microgrids are smart, visualized in apps, knowable in tangible things, embodied in social practices, or easily usable by their owners—only the most interesting ones. Although local electricity generation still seeks its best community apps and identities, the microgrid meme suggests a wish for participation. Thus as new kinds of local energy networks become practical, new kinds of microgrid dashboards must enable them. This too occurs at an intermediate scale, between the handheld device and the vast control room (figure 6.5), and often embedded into physical gear, if not yet into anything worth calling architecture's grid edge. Many kinds of on-site dashboards have gained in use—at first for everyday facility operations and building automation networks, next in energy conservation, soon adding solar panels, and hence next in local energy networks. Without this participatory local network control, the solar panels remain grid tied and are useless in an outage.

Here is another important, if less widely understood, principle of interaction design. Designers sometimes refer to

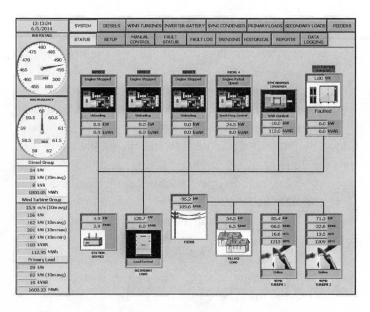

6.5 A microgrid dashboard: larger than a phone app and smaller than a control room. Courtesy of Sustainable Power Systems.

the "indexical" qualities of a form.[16] That word describes how material properties reveal the condition of an object, often though direct cause and effect, like a well-worn path reveals a popular route. Epistemologists refer to "natural data" by which form indicates events, like footprints in the snow. By relying more on such indexical information and thus less on applied instruction, usability engineers seek to reduce the cognitive load. Yet they sometimes forget that human sensibilities are agile enough to discover affordances that have not necessarily been designed and declared. Those are found in the configurations of physical things, such as when a crate serves as a stepladder or where a rock suffices as a table. Thus in the parlance, the habitual usage of an environment heightens the appreciation for indexical signs about changing affordances. Hence a resident is more likely than a visitor to know when to tweak a local system. The intrinsic state of surroundings can itself inform that appreciation.

Casual monitoring seeks some first steps toward that sensibility. For instance, it is a common practice to compare energy usage rate at the moment with the same calendar interval the previous month or year. In one widely successful device, for example, a water fountain keeps count of how many plastic bottles its use has saved. Yet from what ethnographers have learned about interactivity, it would be better that physical environments would themselves bear traces of system status and usage themselves. Whenever being monitored, reconfigured, and affording cyclic patterns of use, places should bear traces of those processes. By use of glow, color, or shape, components of the physical can inform without textual messages. Some of these occur at the scale of an individual plug-in device, some at the scale of the room or the building, and some at the scale of the organization.

Yet so far, most expectation for interactivity has met the local electricity boom via foreground smartphone apps for monitoring and scheduling energy usage (figure 6.6). Although individual residences get most such attention, larger buildings and sites also anticipate new genres of monitoring and participation. This presents a more complex challenge since any such multiuser system would alter conventions socially and would operate in counterpoint to the more usual building operations networks technically, which most users of space never see or think about. Where the engineering goal has typically been to set and forget, now instead it becomes to know, monitor, and collectively tune. Rather than needing to be persuaded by apps, many people are looking for smoother ways to get things done in the sites and situations of everyday life.[17] Because contexts cue activities and help make sense of what is going on, engagement with surroundings can help in, not distract from, getting things done. There is comfort in engaging surroundings that is different from the kind of static physiological comfort that building engineers emphasize.

So now as that expectation changes, new genres of participatory environmental tuning have arisen. Some have been noted here in the context of workplace lighting, for example. In the world of smart home technology, surely the most prominent and genre-making device has been Google Nest. Although studies of Google Nest often show that the novelty soon wears off, and that its role as a data extraction platform remains suspect, the programmability still sells and still does some good, not only for utility bills but also for noticing surroundings in the first place. Somehow the engagement works better than passively inferring habits of spatial usage and comfort. Recent research, however, has had less to say on how comfort remains a pursuit and not a state.[18] Perhaps it is less easy to study, and less countable, how physiological and psychological comfort get conflated,

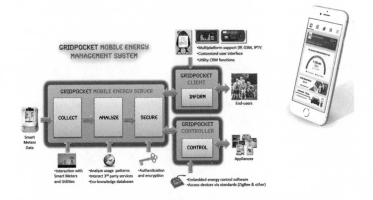

6.6 Convenient mobile access to local energy networks, as shown in an early system diagram (2014) and recent screenshot of a widely admired app. Courtesy of GridPocket.

particularly amid expectations for interactivity. Perhaps it is difficult to engineer for participatory pursuits of contrast and delight. For instance, Google Nest has less to say about someone going over and opening a window simply because the sun is shining.

Meanwhile, despite the exposure of individual smart home devices, the more interesting cultural prospects of course lie in shared social space. Anything that can operate at the community level deserves careful cultural consideration. Anything that can disconnect from the larger net relieves concerns about predatory data extraction. This too suggests some appeal to the microgrid islands—but also for tuning.

The idea of tuning somehow catches a cultural turn. Contrary to past research priorities on predictive modeling, simulation, and verification, it accepts that more adaptive and holistic approaches are also possible today.[19] It recognizes an increasing preference for adaptation and emergence in the incredible complexity of a world flooded with information devices and media.

The idea of tuning nonetheless also turns attention away from those individual devices per se and toward some more ambient, inhabitable landscape of them. That makes more ambient information into a design philosophy of everyday life.

Here again is that fundamental paradox of increasingly responsive technology. In a reversal of twentieth-century aims toward invisible automation, the more that twenty-first-century technology gains sensing and agency, the more that humans now need to keep watch on it, sometimes train it, and expect to tune it.

As interpreted by science, technology, and society writer Richard Coyne, tuning makes places work culturally and not just technically. This is best understood ethnographically. Throughout everyday life, interpersonal relations have to be enacted. Habits matter greatly in this. Habits are acquired inseparably from context. Habits shift when events or interventions make them visible and unfamiliar again. Design seeks to provoke that kind of momentary unfamiliarity and casual social adjustment. It does so less by novelty, or by annotation and instruction, than by embedding into material practice. This makes the Internet of Things culturally significant. For in contrast to being understood as one big system controlled from afar and formulated for predictability in advance, tangible embedding inverts the expectations. It makes local seams and overlaps workable. "The need to calibrate and recalibrate mechanical and electronic devices calls attention to the seams rather than the smooth integration of technology into everyday life," Coyne observed. The many small adjustments at the seams in a world of systems can also fit well into a world of everyday practice. By reopening awareness of so many micropractices, increasingly pervasive media allow, but also demand, more habitual monitoring, tweaking, and tuning. Coyne has recognized this as a medium for attachment to place: "This work attends to the idea of small increments, nudges, and cues, ahead of grand plans and systems. Influences among workers, politicians, and citizens are purveyed most effectively as nudges and subtle shifts in practices that are carried over into technologies, such as pervasive and mobile digital devices."[20]

Thus to counter the less appealing tendencies of a culture of perpetual connective convenience, many people cultivate a sense of conscientious everyday practice. This occurs not only among individuals but also across organizations, and not only in the home but also in shared work and social places. Wherever life demands ever-deepening quantification, monetization, and technological dependency, it also demands digital temperance and tuning. Since so many of those unquantifiable values reside implicitly in everyday life, amid the intrinsic structures of social practices, and so too within the physical situations of built space that frames them, this demands more appreciation of physical surroundings.

Although the smart city in general and smartgrid in particular surely could increase the prospects for tuning, alas too often they have continued to pursue outdated notions of verifiable automation. Instead, as science, technology, and society can help explain, while reciting that a city is only as smart as its citizens, it would be better to let increasingly pervasive media contribute to a richer sense of place. For instance, that could help in the context of more tightly coupled supplies and demands for resource flows like electricity. Again, having more sensate, more locally embedded technologies should afford richer differentiations of conditions. This should improve adaptation to emergent and ever-changing conditions that cannot be predictably formulated in advance. To make more variations more viable therefore involves more human partnership, more willingness to adapt, and more embrace of local difference.

In a practical ethnography of "smart energy technologies in everyday life," interaction designer Yolanda Strengers has examined the importance of tangible participation. With an emphasis on domestic space, this work studied consumption feedback, emissions awareness, time-based pricing, and participatory scheduling or load control. (Larger commercial and industrial spaces, and the organizational policies to go with them, were beyond the scope of this study.) Here the focus is on personal reactions to the smart city imaginarium. So long as energy technologies are perceived as a rational, indisputable way to document and conserve, their social and cultural presence remains unremarkable. Strengers has satirized this worldview with a persona, named "Resource Man."[21] In her critique, Strengers contrasts this "linear and rational model of information exchange" with the kinds of "materially constituted social practices" that more often enact deep cultural value shifts. "By positioning energy-making as a material 'thing' I seek to develop a more nuanced understanding of the role that micro-generated energies play in everyday practice, where energy-as-material meets with constellations of other materials, meanings, and skills," Strengers notes, adding,

Putting this research together, we can infer that energy systems that are materially present, tangible or require some degree of handling, as well as those that are temporally limited, seasonal, intermittent, finite or scarce resources of energy, are better than those that are not. By "better" I mean more likely to enroll household energy making practices that position energy as a tangible, temporal (or seasonal), and limited (or scarce or valuable) material. . . . [Thus] the "success" or otherwise of micro-generation technologies hinges on their ability to reposition the role of energy in practice, or rather their ability to make energies that matter, not a commodity, resource unit or impact, but as a material element of practice.[22]

"Materializing energy" was a series of tangible prototypes developed in the late aughts by interaction designers James Pierce and Eric Paulos. In this work, electronic objects became serendipitous tokens of attachment or "mementos" (figure 6.7). Here was an attempt at personal scale toward energies that matter. How, they asked, might "collecting, keeping, sharing, and activating energy" become subject to the same kinds of personal delights and cultural dispositions that have made serendipitous interactivity into an end in itself for so many other elements of everyday life?[23]

6.7 Energy memento: a small serendipitous token of touching electricity, so to speak.

Responsiveness still needs new genres.[24] So far, across the last decade of the teens, the prospect of ambient interface for local energy practices has been eclipsed by the rush to smartphone apps. For instance, genres of public interactive art that had such imaginative vigor in the aughts have given way to more ordinary, if numerous, festivals of illumination. Open prototypes of physical tokens (phicons) for manipulating information systems have advanced less than closed product designs with embedded information. Voice activation has supplanted graspable activation as the most popular nonvisual interface medium. Yet the prospect of tangible interface remains. Physical presence has become an only more important prospect in local energy systems themselves.

Interacting with Electricity?

In summary, finding any new grid awareness takes some new attitudes and practices. Since electricity has long since disappeared into everyday life, little remains conceived primarily as an energy practice, however. Now that starts changing and gets made temporarily unfamiliar again. Today's renewables, resilience, and on-site network controls begin to reverse some old clichés. People want light and do not understand electricity, said the early electrical infrastructure builders a century ago. Now people want connectivity, do somewhat understand energy, and do want to live more green. Here it may be enough of an advance just to insist on an experiential perspective toward clean local electricity that has otherwise too seldom been considered that way. By now, knowing that the grid is evolving toward participation feels better than imagining it fraying and vulnerable. In a culture that conflates comfort and convenience, that has to mean something. Overall, taking responsibility for resilience seems more convenient than anticipating helplessness. But it does take some awareness of everyday operations. Seeing wind and solar generators nearby invites that anyway. Agency feels better. Feeling ease in all things differs from having everything made easy for you.

Within a new grid awareness, increasingly mediated by the Internet of Things, the question of agency does come up. Microgrids arise from the desire for local control. The Internet of Things demands a new balance between automation and participation. The activation of surroundings is no longer something to ignore, and electric power itself has become something to notice and be glad for again. As sensate fixtures and fittings proliferate at the edges of increasingly modular grids, participants expect a more effortless, occasional sense of engagement and a more reassuring sense of response.[25]

So suffice it to say that "you" are not just all the things done automatically on your behalf. Nor at some seemingly opposite extreme, are you just all the endlessly clickable choices you make.[26] This is the perennial philosophy of human-centered media arts: something has to push back, some kind of engagement must matter, some constraints still must guide engagement, settings cue activities, and habitual contexts still shape sensibilities of practice.[27]

This makes physical scale and configuration only more important. Achieving a well-tuned response makes those configurations more participatory. Or at least that is the cultural expectation when casual interactivity, more so than energy, has so often become an end in itself.

Ethnography sometimes interprets the culture of convenience through what Elizabeth Shove influentially described as the "redefinition of convention, obligation, and normal practice."[28] As a central tenet of social practice theory, almost anything well embedded into everyday life can influence change more than anything imposed from outside. Technological novelties, corporate or government initiatives, or short-lived market fashions seldom match well to deeper customs, on the ground, folded into language, etiquette, and habits—and also into the cues and configurations of physical space. As seen here in the focus on architecture, social practices involve a lot of boundaries and networks in purposefully built space.

So to resume the argument for architecture's role, consider the importance of tangible interfaces (not just handheld touch screens) and situated interactions (not just anytime/anyplace ubiquity). In particular, consider how technology can activate and inform, rather than replace or distract from context. To inhabit an informationally saturated, technologically sensate world can restore, not erode, an appreciation for physical surroundings. Obviously there is more to context than positional coordinates. There is more to background than scenery. The configuration, scale, boundaries, access points, and contact surfaces of physical space inform much about life. Not all information is something sent. To appreciate surroundings need not feel like a luxury, a responsibility, or a burden. Attention to surroundings is not the kind of attention that one must "pay."[29] More effortless engagement with physical circumstances now involves many more elements not only activated but also sensate, not only powered but also sensing, processing, and communicating, and not only visible but also tangible. As an increasingly inhabitable interactivity shapes ever more sensibilities, pursuits of comfort and convenience somehow evolve. As those pursuits grow ever more dependent on electric power but also more appreciative of local difference and more characterized by responsive access to a great variety of networks, a new kind of awareness evolves.

Thus as part of a general move toward a more responsive, resilient, adaptive world, the many engineers, economists, and policy makers who have long overseen the electricity grid must now open up to wider design considerations. Information technology disrupts, and disruptions have long been just what these grid experts have sought to avoid, but perhaps this new era of interpretative flexibility is a positive kind of disruption. Perhaps a wider cultural expectation for responsiveness in more kinds of things can drive interest in the microgrid boom. Digital responsiveness has been an important basis and starting consideration for this writing project. It may yet shape new appreciation for a vital resource that you must not directly touch.

7 TO ISLAND

Resilience drives the microgrid boom, and resilience tends to be local. Locality seldom means isolation, however, so much as difference. Of course resilience is a property of people and places. It is necessary and good that one place adapts differently than another. Locality provides a better scale for participation. Having an everyday stake helps long-term adaptation. Having many resilient locales soon makes for a more adaptable aggregate. Here resumes the argument for urban archipelagos. Contrary to the always-on, all-connecting version of the smart city, resilience puts the focus on clusters. Instead of hardening a full-time network to sustain its form against anticipated disturbances, clusters create a variable network, which may shift its form in unanticipated disturbances. Thus it might matter in unanticipated ways how microgrids can island.

Locality

For good stories of how of communities rise to the occasion amid adversity, one writers' writer to read is Rebecca Solnit. She has become a leading voice of participation. For instance, in her widely read investigations of San Francisco after the great earthquake of 1906 and New Orleans after Hurricane Katrina in 2005, Solnit has told of a "surge of citizenship."[1] Solnit takes interest in how everyday alienation gives way to a more spontaneous resourcefulness—where not obstructed by authority. Of course nobody wants to feel helpless in an emergency. Plainly there is something wrong about lockdowns. Despite the usual lament by the owning class that civilization is just a thin veneer, almost everyone wants to take action.

This is hardly a fringe position. "Social connectedness and a culture of mutual help have a major outcome on the impact of disasters," observes the United Nations' Making Cities Resilient campaign (figure 7.1).[2] Greater resilience arises where social practices of trust and participation already exist. Today a wide spectrum of city governments, relief organizations, environmental activists, and climate justice campaigns works on readiness; everyday citizen participation in such organizations prepares for resourcefulness in emergencies. According to the Community and Regional Resilience Institute, a project of the Meridian Institute, a prominent think tank that facilitates decision making in hybrid forums, resilience resides in an everyday mix of community services. "Resilience is an inherent and dynamic attribute of the community. This means that it exists throughout the life of the community."[3]

For example, after Superstorm Sandy, the city of New York worked with the Urban Land Institute, an influential organization of design professionals for public education, to develop a set of over twenty policy recommendations toward improving local resilience. Among these recommendations: cities must acknowledge local difference in infrastructures— that is, "identify local land use typologies; allow for safe failure on some noncritical infrastructure systems; [and] devolve funding to the lowest effective level where appropriate."[4]

DISASTER SUPPORT HUB LOCATIONS

To learn more about what you can do to prepare for emergencies, visit: **vancouver.ca/beprepared**

CITY OF VANCOUVER

LOCATIONS WILL FEATURE THIS SIGN

DISASTER SUPPORT HUB

Disaster Support Hubs are located at 25 sites throughout Vancouver to serve as public areas where citizens can gather following an earthquake or other natural disaster to share information and resources.

1 Britannia Community Services Centre
1661 Napier Street

2 Champlain Heights Community Centre
3350 Maquinna Drive

3 Coal Harbour Community Centre
480 Broughton Street

4 Creekside Community Recreation Centre
1 Athletes Way

5 Douglas Park Community Centre
801 West 22nd Avenue

6 Dunbar Community Centre
4747 Dunbar Street

7 False Creek Community Centre
1318 Cartwright Street

8 Fraserview Branch – Vancouver Public Library
1950 Argyle Drive

9 Hastings Community Centre
3096 East Hastings Street

10 Hillcrest Centre
4575 Clancy Loranger Way

11 Kensington Community Centre
5175 Dumfries Street

12 Kerrisdale Community Centre
5851 West Boulevard

13 Killarney Community Centre
6260 Killarney Street

14 Kitsilano War Memorial Community Centre
2690 Larch Street

15 Marpole-Oakridge Community Centre
990 West 59th Avenue

16 Mount Pleasant Community Centre
1 Kingsway

17 Oppenheimer Park
400 Powell Street

18 Renfrew Park Community Centre
2929 East 22nd Avenue

19 Roundhouse Community Arts and Recreation Centre
181 Roundhouse Mews

20 Strathcona Community Centre
601 Keefer Street

21 Sunset Community Centre
6810 Main Street

22 Thunderbird Community Centre
2311 Cassiar Street

23 Trout Lake Community Centre
3360 Victoria Drive

24 West End Community Centre
870 Denman Street

25 West Point Grey Community Centre
4397 West 2nd Avenue

7.1 Participatory response: a map of Vancouver's archipelago of disaster support hubs. Courtesy of Vancouver.ca.

In North America, where state and local governments have taken the lead on clean energy strategies, often that level is the city government. While state governments can set major policies like renewable portfolio standards, cities do better at integration across related concerns in, say, education, mobility, and land use. This is a rising opportunity and challenge. Thus the 100 Resilient Cities project of the Rockefeller Foundation recommends for cities each to appoint a chief resilience officer.

In a public education piece, "Six Foundations for Building Community Resilience," the Post Carbon Institute explains: "When we intervene in a system with the aim of building its resilience, we are intentionally guiding the process of adaptation in an attempt to preserve some qualities and to allow others to fade away—all while retaining the essential nature, or 'identity,' of the system. Thus, resilience building necessarily starts with decisions about what we value."[5]

Of course "community" is not just physical proximity, nor market segment, nor a unit of real estate development but also a set of values. While almost anyone might have their own basis of belonging, a true community almost always shares some resources, goals, and protocols, besides physical space. Interpersonal patterns often shape the understanding of place. The direct users of a resource often can manage it better than anyone else.

As the Community and Regional Resilience Institute observes, "Any adaptation must improve the community, i.e., must result in a positive outcome (positive trajectory) for the community relative to its state after experiencing adversity."[6] Resilience is not about avoiding change. Whereas sustainability might simply hope to bounce back, local adaption can sometimes "bounce forward." Although that expression has mainly been popularized in the context of personal resilience, it can also express useful properties of complex adaptive systems. As the Detroit-based Kresge Foundation has explained, resilience "conceived simply as 'bouncing back'" not only can retrench existing problems but could also "be co-opted by opponents of meaningful reform."[7]

Clusters

So to resume the argument on microgrid institutions, there must be some appeal at community scale. Where private backups might create new social divisions, shared resilience might create new institutions. To value adaptation is to value recombinant change. Systems make internal changes to keep going. Clusters tend to do well at that.

Although community energy choice has gained already strong appeal for being clean and green, it can also bring better resilience. The Clean Coalition argues this well: once the community and not just individual sites' backups becomes the scale of a microgrid, it becomes easier to prioritize sites, stage everyday trading, expand coverage areas, and of course create local investment and jobs. More so than private sites adapting themselves one roof at a time, this creates a pattern that is easier to replicate, to interconnect, and to cluster.[8]

To understand this in a bit more detail, briefly consider some systems thinking. To begin, it helps to recite a first principle of ecology that no system can be explained through focusing on a single scale alone, or from the perspective of a single discipline alone. Hence as the Post Carbon Institute, for one, advises, "Recognizing that there is more than one way to see things is at the heart of systems thinking."[9] So here recall (from the earlier argument on smart green blues) the science, technology, and society concept of a "technological frame." That expression describes how the established bias of any one discipline can preempt other outlooks. For instance, it is often said that public utilities do not so much oppose change as that they are just not built to comprehend it.[10] So then to repeat from the earlier argument, whenever separate well-established fields cannot each resolve a challenge alone, some "triangulation" from outside their technological frames can help. Sometimes, helpful perspectives come from new points on the edges.

To the contrary, the established bias of each respective frame can preempt other outlooks that would be useful in case of disruptions. Under a given frame, a system then gets well engineered for anticipated conditions, yet may respond poorly to unanticipated conditions. Layers of complexity (and disciplinary monoculture) may add to predictable resilience but create many small triggers of unpredictable problems. Systems thinkers often describe such phenomena as "robust yet fragile."

According to cluster theory, looser networks of simpler entities may do better. To become more resilient, a network must be variable enough to reconfigure itself in response to external shock. In an oft-cited work on cluster systems ecology, resilience strategist Andrew Zolli and journalist Ann Marie Healy have investigated this "balancing act" as a way of getting past robust yet fragile systems. Across examples from forestry, immunology, marine biology, and the internet, Zolli and Healy have explained how clusters that are "dense, diverse, distributed" adapt better than their more monocultural counterparts. In urbanism, such effects occur at both the micro scale of neighborhood public health and safety, and the macro scale of regional competitive advantage. Cities have always afforded greater variations than rural life, and today this advantage is sufficient in itself to drive unprecedented rapid urbanization. Despite their great size, though, some regions embrace change better than others. To Zolli and Healy, "The answer is not in their scale, but in their clustering of diversity and density."[11]

Many such explanations trace their argument to a standard of systems literature in which ecologists Lance Gunderson and C. S. Holling have explained nonuniform clustering by means of a term adapted from nineteenth-century utopian political theory: *panarchy*.[12] Whereas the more familiar word "anarchy" describes an absence of rule, panarchy depicts a condition of rule by all or from all levels. As applied to ecology, this captures how adaptive cycles interweave, especially where nested in a hierarchy of scale. Larger, slower cycles not only contain smaller, faster ones but can also themselves be changed by those. A development or disturbance within one cycle might affect the configuration of a larger one. In a panarchy, effects are two way, intermittent, and capable of large-scale reconfiguration. Interconnections are not so much continuous as episodic.

Ultimately no one system can anticipate or interlink everything. The greater the number and diversity of things at the edge of a system, the more the system needs intermediate and variable layers. The better those things can reorganize as necessary on local layers, the less they burden anything at the center or the top. The more things are prioritized from the edges, the more you might expect much else also to change.

From still earlier in the opening argument here, recall the distinction made by Clifford Siskin between living as a helpless subject of "the system" and living as a participant in a "world of systems."[13] Here in the context of resilient panarchy, that distinction takes on additional meaning. And in the context of more resilient local energy, that meaning suggests a better layer model, with better differentiation of interconnected islands.

Islands

Clusters of specialized islands occur not only in systems thinking but also in some larger cultural imagination. From mythological Atlantis to reality television, island poetics have shaped many kinds of thinking. It has been said that Earth itself is an archipelago, and that life itself is an island in the vastness of space. To designers, islands appeal not only in their bounds but also their metaphors.[14] To the influential philosopher Immanuel Kant, in one of his most cited passages, all that is knowable ("the territory of pure understanding") is like an island. "This domain is an island—seductive name!—surrounded by a wide and stormy ocean, the native home of illusion, where many a fog bank and many a swiftly melting iceberg give the deceptive appearance of farther shores, deluding the adventurous seafarer ever anew with empty hopes."[15]

Utopia itself was an island in Thomas More's famous book of 1516 (figure 7.2). Literally "no-place" yet set in the new world of the Western Hemisphere, this work epitomizes the use of an island as a symbol of a wish to start anew. Literature abounds with new starts on faraway islands, where narratives get taken to greater extremes, good or bad, than would be conceivable in the established contexts of everyday life. To adventure among archipelagos as Odysseus and Gulliver each so famously did, likewise appeals to the imagination. Such a poetic apparently serves a deep human need for one place to be different from another.

In a recent work named *Islandology*, literary critic Marc Shell has sounded this perennial social construct. Today island studies is an academic discipline. Cultural geographers examine what happens to islands when bridges connect, tourists arrive, species invade, or waters rise. Political philosophers ask to what extent any person, state, or cultural preference is or is not itself an island. Urbanists examine how clear boundaries afford more daring designs. To Shell, this bounding effect seems most vital: islands are above all distinct. "Islandology cannot afford to avoid defining definition," he begins his book.[16] More so than most other ideas or entities, an island provides an especially clear domain for unique sensibilities to occur. Not only are differences of practice more distinct, but so too is the tendency to identify with them. On an island, to follow the shore is to return soon

7.2 Utopia was an island: woodcut from seventeenth-century edition of Thomas More's *Utopia* (1614). Wikimedia Commons.

enough to your starting point. Civilization grew not so much on the bodies of continents as around the edges of oceans. Here too the interesting parts are out at the edges.

Yet to identify with perimeter may emphasize security as well. The island is also the nation-state, where in everyday politics, much as Kant saw for knowledge, one should never trust others from across the water. To island may be to give up on the others. When enough parties retreat inside their respective perimeters, the remaining shared space is suspect and rogue. Conversely, islands can go rogue, and privileged rogues can take their action offshore to islands.

How does the cultural resonance of island metaphors help explain the appeal of the microgrid meme? Freedom to island has resonance in an era where perpetual connectivity has otherwise assumed some dystopian tones. Clearly there is more to local electricity than grid modernization. Clusters of islands matter in that; a network in which everyone is an isolated perpetual user of a large distant infrastructure might not be as adaptable as one made of recombinant clusters. For above all, participation again counts in resources that were formerly made universal and invisible. A grid that affords no participation might not be smart at all, especially not in the face of increasing disruptions. As noted for politics, technology, and urbanism alike, locality has become the most workable scale for energy democracy.

Yet that same freedom to island appeals to neoliberal outlooks that hold just about anything public in disgrace. Much like the gated residential subdivisions in, say, Houston or Los Angeles, and as demonstrated to the extreme in the recent troubles over rebuilding Puerto Rico, private local energy can itself assume dystopian tones. This is perhaps the very worst thing one can say about microgrids as backup: they create an important new category of haves and have-nots.

This is all the more reason to emphasize the microgrid meme at the community scale and the archipelago, and for its emergent advantages. Remember too that as an off-grid island brings electricity more quickly to world's billion people who never had it, the microgrid meme itself plays differently in different places.

There is, however, a more positive poetic at work. There is more to microgrids' coolness than having backup power installed. This, of course, has to do with clustering. Whereas an island as starting over would be more like an ecovillage, or an island as turning its back on the world would be more like a doomsday preppers' camp, and both of those might go permanently off grid, an island as a distinctive member of an archipelago might be something better. It might value intermittent or episodic connections. It might connect not only with some larger grid but also among many other such islands in its cluster.

So wherever there are islands, ask how they cluster and interlink. Ask how their relations allow each to become more distinct. Ask how a clear boundary not only of form but also of practice intensifies their designs. Already the more interesting microgrids are the grid-connected ones. Does enough density of them eventually turn the proprieties of the grid upside down? It may be too early to tell, but not too early, per Shell's many political insights, to rethink island thinking.

Grid(s)

If something about islands explains the appeal of the microgrid all beyond its immediate feasibility, then when combined with something about ecological cluster theory, this appeal quickly morphs into fantasies of a bottom-up infrastructure revolution. After all, even the giant energy companies admit of one. So before taking up the perspective of urbanism on that, as too little work in science, technology, and society has done, and with the belief that a new grid awareness belongs to the rest of us too, consider one more brief glimpse at that vast machine.

Since it is not the role of this project to offer a bright green future (nor to weep for a darker one) but only to interpret the microgrid meme as a catalyst for a welcome new grid awareness, here allow some last few disclaimers. To qualify today's enthusiasm for cluster ecologies and perhaps too an archetypal wish for islands, it is important to recognize what the community microgrid can and cannot provide. Lest this seem too much a jump from island poetics or systems ecology, consider its relevance to three basic grid awareness arguments already in play.

First and foremost, big is good here. There is reason to repeat that the grid is the world's largest machine, and for all its issues, this is still something to marvel about. This is the main engineering argument against lonely rogue islands: the more resources on a network, the easier it is to swap in another when any one goes out. Although the economies of scale in power generation have reversed in recent decades, those for reliability and balancing have not. This network has enormous physical inertia. Its foremost priority is in the physics of maintaining power quality. Already this might seem like more detailed grid awareness than almost anyone ever cares to have, even for a moment, but the point is that more people need to understand this now. Dare to imagine the physical magnitude of several gigawatts of energy embodied as carefully synched waveforms on a regionally scaled network of high-voltage transmission lines. Besides all the dynamos and windmills that drive it, this too has inertia. This is what those giant control rooms manage in real time as supplies and demands fluctuate through the day. There, whenever some major resource swaps in and out, that enormous inertia must stabilize within minutes and sometimes seconds. It does so in the domain of voltage frequency. If you can understand that at all, say some experts, you have begun to understand the grid.[17] So if a large source of supply or zone of demand suddenly goes down, try to imagine how many things have to synch to get a region quickly back up to a steady 60 hertz from a momentary lull down to, say, 59.5. In this geeky reality, there is not just engineering but also politics. Indeed here is the technological frame (i.e. preexisting set of assumptions) of the public utility company: controlling all the edges really helps maintain the physical inertia of the entire system.

In the second main argument here, alas the modern electric grid never acquired a strong layer model, not like telecommunications has done, for instance. This is largely for the reasons above, but also since so much of the technology predates most information technology protocols. Almost everything still runs from that one transmission layer, run by those regional operators. To understand how that may or may not change, this is where the language of systems ecology can come in. Although of course variably dispatched, electric power never developed much panarchy: it was always one way, always on, with relatively few players, and rigidly configured. The metaphor of water pools held well, for the flow was always in one direction, from the top, with the big transmission lines as the sluices and gates. That reality has little capacity for two-way flow, however, nor for intermittent islands. Thus for quite a long time, the few big transmission operators have remained the one layer to which all else must connect, and the only layer at which transfers between regional pools can occur. This deserves wonder for remaining so robust, but it also deserves concern for becoming more fragile.

Third, today's major trends toward decentralization invite many more fields to find their roles at the edges. When these trends disturb the big-is-good technological frame of the power industry, that can lead to smart green blues. Informational superabundance drives this. The weakly layered existing model has little capacity for handling the explosive growth of information traffic to and from so many more entities now at grid edge. As the number of players goes from dozens to thousands, and eventually millions, eventually no one layer can keep track of it all. But it is also becoming important for the physical energy itself. Even more so than the data, there is no easy way for all that energy to flow back up the hierarchy. Today when it is everyday news that too many renewables threaten the top-down stability and constituency of the grid, the inevitable alternatives work from the bottom up, and intermittently.

Meanwhile, as again asserted especially well by the Clean Coalition, which is arguably the most effective intermediary between the utilities and their publics for advancing community microgrids, many advantages are neither so much top down nor bottom up as increasingly *lateral*. The fact is that ever more electricity gets locally produced and consumed without ever going onto the regional transmission grid at all. When islands can interlink, this creates new markets among local layers of distribution rather than all at the regional transmission level as before. As the Clean Coalition explains, community-scale microgrids can be more easily staged than private backup ones. That steps up returns on the local investment: "[building a] wholesale distribution grid unleashes renewables." But this does require better layers and better islands.

Thus the abstract prospect for an archipelago should be obvious enough and appealing not only in its bounds but also its metaphors. A lateral network of real time transactions and two-way distributions can improve (not detract from) the reliability and quality of something larger than themselves. Enough of that is apparent in complexity simulations, but sometimes the speculations go further. A scenario exists (counterfactual to today's realities of power reliability, quality, and pricing) where everyone is on at least some sort of microgrid, and the legacy macrogrid remains mainly as a networking platform, for occasional transmission and trading, and as an expensive backup. But so far that is just speculation.

Yet the need for a better new layer model is understandable enough. For instance, interpreting engineering, policy, and economic debates, and with inspiration from European practices that are ahead on some of this, eminent journalist David Roberts has characterized a new kind of layered distribution organization (figure 7.3). Any given layer can

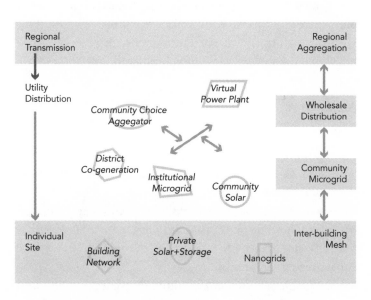

7.3 Toward a grid future with a deeper layer model, clustered islands, and multidirectional flow.

communicate both up and down, and in doing so, reduces the number of communications to more manageable volumes and formats. Although the power quality engineering and redundant resource economics run against established practices, it may be unwise to assume that must always be so. If ever the local layers at the bottom can attend to their own use optimization even before making any exchanges with those above, then the entire industry does indeed turn upside down. "Big, centralized power plants become the last resort, not the first." As Roberts speculates, "What flips is the priority, and with it, the power. Foregrounding local resources would at long last make cities and regions (their vehicle fleets, their building and zoning codes, their infrastructure, their vulnerabilities) full partners in optimizing and decarbonizing energy."[18] Here is a fresh kind of argument for urbanism.

Remember, resilience tends to be local. However it plays, here is a "massive cultural system" turning toward the highest new priority of the times, and with it bringing the physical environment back into consideration.[19]

Archipelago

Within that open conversation, remember the city, question the smart city, and ask the meaning of islands. For a single word to take away from this reading, it appears timely to renew the islands metaphor of *archipelago*.

For one last specific example here, consider indicators of an archipelago in Boston, in what was apparently the first citywide study of its kind. In the mid-2010s, several collaborating organizations developed a recommendation for approximately two dozen community energy zones most pertinent to Boston's overall urban energy resilience (figure 7.4). Starting from existing city data for building types, sizes,

Selected Analysis Zones in Boston

Multiuser Microgrid

Energy Justice Microgrid

Emergency Microgrid

7.4 The urban archipelago: a map of the twenty-two most suitable microgrid sites identified by a spatial analysis of building types and energy uses. Courtesy of MIT Lincoln Laboratory.

and configurations, the MIT Lincoln Laboratory simulated hourly thermal and electrical demand at higher spatial resolution than any known precedent had achieved.[20] This mapped not only high-energy-intensity areas but also critical uses, which together helped identify potential "anchor buildings" for local energy infrastructure investment. Collaborators at the MIT Lincoln Laboratory developed a process for interpreting this data set via factors of emergency preparedness, social welfare, and anchor institutions. An archipelago of institutional and campus sites figured clearly in the results. For each such site, and using the feasibility model from Berkeley Lab (where the word "microgrid" first caught on), the researchers developed a recommendation on the energy mix and microgrid integration strategy.[21] When combined with the building type and social resilience data, this made for a substantial demonstration of urban "big data" in action. The results indicated an archipelago of potential resilience nodes, many of them institutions. Because the city is always an aggregate and not a continuum, this also suggests how many agendas of resilience occur in the context of urban design.

Island thinking does help explain why work from the perspective of architecture and urban design might take up the microgrid meme. As these disciplines often understand best, there are always cities within cities. No amount of flux by broadcast media, automobile culture, or internet apps can dissolve that. The universal net has not negated local geometry. To do anything anytime anyplace would be nowhere. That's not what the high rents indicate is happening. Instead, the particular differences among pieces of the urban aggregate regain cultural value. Compounds, districts, campuses, neighborhoods, estates, monuments, and megastructures, each with a unique configuration, fixed circumstances, and at least partial autonomy, together comprise a much richer environment than one where anything might happen anywhere.

Architecture and urbanism learned this sooner than some other disciplines, and learned it the hard way. Back in the twentieth century, the modernist attempt to create a fully transparent, mobile, rational city lost sight of the aggregate, with dreary results. Midcentury programs of behavioral problem solving, neighborhood demolition, automobile monoculture, transparent space, rational form, and standardized uniformity all generally favored the continuum. In the rational language of mathematics, they sought to make space more "isotropic": the same, or varying at the same rate, in any and all directions. Yet the more that central planning imposed such a continuum, the worse the cities became. That tragic history should be better known to anyone who would repeat such category errors in any rush to build the smart city.[22]

One famous moment of counterreaction deserves retelling in the context of island thinking. In 1977, German architect Oswald Matias Ungers, an early critic of the modernist continuum, collaborated with a young Rem Koolhaas, who went on to become the most influential voice in architecture, on a short (six-page) manifesto named "The City in the City—Berlin: A Green Archipelago."[23] Thanks to a retrospective reissue in 2013, this manifesto has regained exposure today.[24]

Like many cities in that newly postindustrial era, Berlin was depopulating. Although an exceptional case since closed off by the Berlin Wall, it foretold issues that soon arose elsewhere too. The design disciplines, still focused on modernist worldmaking, as yet lacked ways to address cities that were now shrinking. Without sufficient prosperity to address spatial issues citywide, postindustrial cities now had to learn how to choose their focus—how to design interventions in limited areas that could still catalyze larger areas. The way to revive Berlin's isolated western half, this manifesto recognized, was to concentrate on the many separate, relatively solid, and deliberately exceptional islands. Each of these would provide a microcosm of legible civics amid the postindustrial decline—hence "the city in the city."

Today this poetic of the archipelago conversely gains significance in counterpoint to indiscriminate hypergrowth too, which is the world's more usual urban condition. As architecture critic Pier Vittorio Aureli has articulated in long historic perspective, a civilization must work to uphold political participation (which the ancient Romans called *civitas*) against the more banal, private, and often solely economic proliferation (which the Romans called *urbs*).

Alas, the modern economic city is almost all urbs, or what by the millennium Koolhaas notoriously called "junkspace." Of course neoliberal market ideology is itself a form of politics, but the point is that once computers became commonplace, finance began to rule all else. As Aureli observed, some categorical error in all this proliferation could be traced as early as the nineteenth century, in what philosopher Georg Wilhelm Friedrich Hegel characterized as "bad infinity," and what, toward an island poetic, urban planner Ildefonso Cerdà lamented (sounding a bit like Kant) as a "vast swirling ocean of persons, of things, of interests of every sort, of a thousand diverse elements."[25] Where once commerce was a means to a civilization, by now it has become an end in itself: an ocean of urbs that erodes all else. Thus in perhaps his most cited remark, Aureli described the sea that the archipelago must exist within, but resist: "The essence of urbanization is therefore the destruction of any limit, boundary, or form that is not the infinite, compulsive repetition of its own reproduction."[26] Perhaps as so many critics since Hegel have said, modernity's endless replications are compulsive precisely because so categorically flawed.

So then to uphold any civitas, in counterpoint to endless urbs, architecture must exist as archipelagos. There the filtering, embellishment, and intermittency of boundary crossings becomes its most basic operation. After all, architecture's walls and openings have done so for thousands of years. To Aureli, "If one were to summarize life in a city and life in a building in one gesture, it would have to be that of passing through borders. Every moment of our existence is a continuous movement through space defined by walls."[27] To close a courtyard gate is to island.

Remember that islands can go rogue, however. To island is to turn one's back on the others, to identify with us against them, to abandon notions of improving any larger sphere, and so in sum, to identify with perimeter mainly as security. Thus the main critique of Aureli's position concerns its origins in xenophobia and islands' tendencies toward divisiveness.

So to temper the argument for urban archipelagos (much as for microgrid galaxies), likewise consider three qualifications. First, not only architecture but likewise many other disciplines work with both objects and networks, as if in oscillation between the figure and the field. Designers who recognize today's preeminence of networked flows nevertheless create objects: the fixtures and fittings to make those flows more knowable and usable. Second, as is worth repeating frequently here, it is not the individual urban islands that disconnect defiantly but instead the ones that interconnect intermittently, often quite elegantly, that turn out to be the interesting ones. Third, in the overall imagination (and at last bringing power to more remote parts of the world), the island is as often good as bad. Moreover as More showed with *Utopia*, the island remains a strangely effective means of cultural critique. Again, the boundedness of an island allows more intensive design experiments, not only on the ground, but also in the outlooks.

Imaginarium

The pertinence of this theoretical episode to the big business of *the smart city* is this: question continuum. The compulsion to connect exactly everything may be just the latest stage in some big, long-term category error. If today the smart city seems like some especially flawed infinity, then these architects' thoughts on archipelagos seem worth remembering. Thus besides serving as a catalyst to local energy democracy and a new grid awareness, the idea of a microgrid could well prompt some rethinking of the smart city party line. Question mandatory connectivity. Watch out for rogue cities within the city. Yet imagine that even the smart city might have worthwhile archipelagos. Since so much of this concerns the usability of complex places, call it an *imaginarium* (figure 7.5).

Practical urban informatics do exist. This has become a very real business sector for the likes of Cisco, Siemens, Microsoft, or IBM.[28] Better sensing, processing, and communications do help manage quite a range of city services, often with far better interconnections than was conceivable before, and enough perhaps to be called smart. At the street level, in the accessibility of public records, and in many aspects of everyday service integration, it all seems real enough.

Today as city councils invest billions into urban information systems, though, it seems vital to recite that the complexities of the world cannot be reduced to proprietary engineering solutions. Information infrastructure builders might have the best insights of the moment, but not the only ones, and not into everything. The issues that urban sociologists, architects, political activists, and neighborhood associations have been working on for lifetimes cannot be reduced to data streams, may not be quantifiable, and should not be forgotten.

As must always be said, a city is only as smart as its citizens. In an early event on Smart Citizens (2013), respected urban futurists Drew Hemment and Anthony Townsend asserted the importance of local difference, noting that "the value of bringing citizens into the process is that only they can turn cookie-cutter corporate plans for the Smart City into designs that are truly bespoke." Well within the spirit of Jane Jacobs, Hemment and Townsend anticipated today's emphasis on clusters: "Alongside 'top-down' master-planning, we need to enable 'bottom-up' innovation and collaborative ways of developing systems out of many, loosely joined parts."[29]

7.5 Stanza, *The Nemesis Machine* (2015–2019), a real-time, data-driven representation of urban complexity. Courtesy of the artist.

Much as per architect Christopher Alexander's famous essay "A City Is Not a Tree," in which he cautioned how "there is some essential ingredient missing from artificial cities," likewise, urban informatics critic Shannon Mattern asserts that "A City Is Not a Computer."[30] Too much about life is socially and physically enacted ever to be transmitted. "We, humans, make urban information by various means: through sensory experience, through long-term exposure to a place, and, yes, by systematically filtering data. It's essential to make space in our cities for those diverse methods of knowledge production."[31] As if an archipelago, each chunk of social infrastructure must embrace spatial limits, celebrate local difference, and intensify distinctions from other islands.

Stop saying "smart cities," futurist Bruce Sterling advised a larger mainstream audience in *The Atlantic* in 2017. "Digital stardust won't magically make future cities more affordable or resilient." Instead, the expression is more a shibboleth of boosterism, or as Sterling quipped, "The language of Smart City is always Global Business English, no matter what town you're in."[32] Still this technocratic tribe needs a name. There is reason to admit how better sensing, processing, and communications do help manage quite a range of urban resource networks, often with far better interconnections than was conceivable before. Yet so much as it is comprised of systems, the city is a world of systems, not one system. Hence this writing has used *the smart city*, always with the definite article in the singular, to emphasize that. How different that imagination than one that thinks about islands.

To call it an *imaginarium* reveals something more elusive, however. Culturally, the smart city is neither a utopian fantasy nor a mundane business sector so much as a name for everything that has yet to be implemented. Whatever has already been achieved isn't it. As the cycle of hype and disappointment alas so often demonstrates, the prospect proves ever more interesting than the realization.

So question totality and consider intermittency.[33] Imagine, as architects have done, clusters of cities within a city, each of which makes different choices about connectivity, and any of which has the means to disconnect occasionally, to reconfigure connections adaptively, and in doing so, to create cultural patterns.

To speak of archipelagos counters the recent obsession with perpetual connectivity. Now that the social, political, and environmental costs of smartphone obsessions start to become recognizable, it becomes fair to ask for intermittency, and not only the means but also the rights and occasional advantages to intermittency. In this, note some similarity to obsessions with automobiles half a century ago, which likewise long went unquestioned. Where not so long ago automobile traffic calming would have been heresy, now that has become proper urbanism. Today's urbanists know to get people out of their cars. Today nothing raises property values like walkability. Likewise, where not so long ago, disconnecting from the internet for so much as a hour would have been preposterous, today it is good mindfulness. Digital temperance has become necessary. Nothing raises respect like undivided personal presence.

So when microgrids enable many more kinds of urban entities to go on and off a larger network voluntarily, this is not just about electricity. *To island* has become a necessary tactic against prospects of mandatory connectivity in other aspects of life as well.

Downtime

The next time your city suffers a superstorm, many more zones can stay powered. Neighborhoods, development districts, ecoblocks, and community solar schemes have seen to that. Yet microgrids are not just for emergency backups, and island intermittency is not just for outages. A more fundamental shift is in play.

To island invokes many vivid notions seen here: clusters, intermittency, security, waiting on standby, going off grid, running naturally or at zero carbon, and sometimes not needing to run anything at all. Any noun that has been verbed is hardly the best of words, however. Since the prospects of outage drive so much of this, and above all since the prospect of disaster conflates and confuses just about everything, this project has appropriated a different word to represent this new state of mind. Just call it *downtime*.

That word normally describes a state of system that is temporarily unavailable, whether from unexpected fault or scheduled upkeep. When used socially, though, it also describes voluntary disconnection. Philosophically, it suggests that ever necessary but ever more elusive state of repose. There is nothing wrong with that. There is nothing utopian about it, and there are some always-on (or suddenly all-off) dystopias it may help avoid.

The ability to set up local boundaries does not mean a wish to keep out everything else. The capacity to disconnect from one system does not mean being cut off from all the others. It does not mean going off grid permanently. It does not pretend that the rest of the world does not exist. It does not involve defiant sacrifices. In the case of resilient local electricity, the capacity to disconnect sometimes does not even mean going without but *just going with one's own.*

So let "downtime on the microgrid" take on multiple meanings: not only involuntary emergencies but also the practice of intermittent connectivity within an archipelago, and within the latter, not so much the process of tuning out as a new way of tuning in. Let this seem less like shutting down and more like standing by. Whether on bad days of emergency, good days when the sun is out and the wind is blowing, or just according to ongoing fluctuations of local and regional grid dispatch markets, this more versatile notion of downtime has its moments. As if resilience involves personal, social, and not only technological readiness, do seek ways to recall how life can carry on (and until the last century, always did so) without full-time powered connectivity.

Within the agendas of local electricity, let this notion admit a deeper need for priorities. Whereas the grid has been engineered to provide the same high reliability everywhere, all the time, for all uses, whatever the environmental cost, in fact some uses are more critical than others, or need more constant, high-quality power than others. Let that illustrate much else too. Whereas anytime/anyplace convenience has become a cultural obsession, in fact some things are better done in some places and not others, at some times and not others, and sometimes not at all. To have it all anywhere is some kind of nowhere. To have everything always on, infinitely interconnected by an end-to-end, all-seeing smartgrid, by now appears at least as a cultural fallacy and at worst as a cultural dystopia. For eventually it has an outage. Despite all the best efforts at hardening, it still seems robust but fragile. All of it could conceivably fault at once.

"Cities made of atoms and data bits sometimes seem more fragile than their predecessors built of brick, stone, and concrete," urban technology historian Antoine Picon has observed. "Electricity remains the city's principle source of weakness: it is no accident that so many futuristic novels depict its disappearance as a precursor to the apocalypse or to a return to barbarity."[34] For example, in Emily St. John Mandel's *Station Eleven*, which is thought to be one of the few refined literary instances in the untamed postapocalyptic genre, the plot resolves with a few hardy wandering survivors sighting "in the distance, pinpricks of light arranged into a grid."[35]

Somehow the smart city imaginarium rarely mentions fragility. Perhaps it imagines that panoptic oversight will prevent anything bad from ever happening. Yet the more immediate reality is unprecedented storms. Perhaps future sensate systems will speed every recovery. A proliferation of information devices alas also proliferates cybersecurity risk. Under some higher truth, wisely planned readiness cannot be readiness for everything.

The critique of totality in the smart city imaginarium helps highlight that condition. Early developments in microgrid institutions help suggest complementary thinking. To repeat in refrain, laments on technology don't get far. Yet celebration and needless deepening of technological dependency has little wisdom either. Already the more sensible resilience strategy trades some universality for locality. Simply modernizing last century's infrastructures, cultural beliefs, and building patterns may no longer remain an option anyway. Values must change. Much of the cultural shift concerns just what needs to be always on, what might tolerate intermittency, and what might even benefit from some downtime.

Intermittency in at least something somehow becomes socially acceptable. Intermittency might restore some sense of wonder that should rightly surround a resource that can no longer be taken for granted. The belief that almost everyone can ignore electricity may no longer hold. Like João Penalva photographing lines above the streets of Osaka (figure 7.6), you might want to find some new grid awareness.

In closing refrain, freedom to island has resonance in an era where perpetual connectivity has otherwise assumed some dystopian tones. As a way to push back against the more monopolistic aspects of technofuturism, there is nothing like capacity to disconnect sometimes. To the distant digital overlords in their cloud, disconnection is heresy. Although both the internet and the grid that powers it need to remain up continuously, like a heart, nevertheless its participants and contexts of use do need some independence and rest, more like a brain. This may become the parting thought in much else too. To live in such a vulnerable world takes some new readiness to let go of *something* . . .

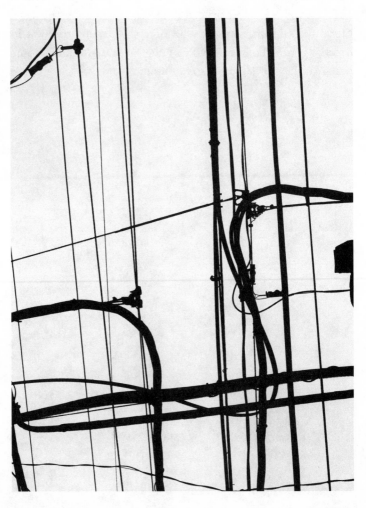

7.6 João Penalva, *Looking Up in Osaka*, Ebisu Higashi cho-me, *Naniiwa-ku* #2 (2005–2006). Courtesy of the artist and Simon Lee Gallery.

NOTES

1 At the Edge

1. "What would you miss?," the first line here, likewise appears as the leader in the press endleaves for the paperback edition of novelist Emily St. John Mandel's *Station Eleven* (New York: Alfred A. Knopf, 2014). Yet it is too inclusive a question to credit to any one source. Meanwhile, "the vulnerable world of the Anthropocene" is best summarized in Roy Scranton's terse, remarkably readable *Learning to Die in the Anthropocene: Reflections on the End of a Civilization* (San Francisco: City Lights Books, 2015). Then that's it for the dystopian keynote! *Downtime on the Microgrid* is not a work of dystopian doom and gloom, nor conversely is it technoutopian. In an era and nation ever less prepared for ambiguity and intermittency, this book seeks both of those, and thus a middle ground.

2. "Watching television by candlelight" is attributed to the stand-up comedian George Gobel, in a televised tribute to Thomas Edison, by the prominent producer David O. Selznick, in 1954.

3. See Andrew Zolli and Anne Marie Healy, *Resilience: Why Things Bounce Back* (New York: Simon and Schuster, 2012), 93–117.

4. The term "grid edge" was coined by Greentech Media in 2013. See Mike Munsell, "Greentech Media Defines the Future of the Electricity System, and It's Called Grid Edge," Greentech Media, October 7, 2013, https://www.greentechmedia.com/articles/read/greentech-media-defines-the-future-of-the-electricity-system-and-its-called#gs.40x4an.

5. Clifford Siskin, *System: The Shaping of Modern Knowledge* (Cambridge, MA: MIT Press, 2016), 3–21.

6. Tanja Winther, *The Impact of Electricity: Development, Desires and Dilemmas* (New York: Berghan, 2008), 217.

7. Alan D. Pasternak, "The United Nations' Human Development Index and Electricity Use, 60 Countries, 1997," in *Global Energy Futures and Human Development: A Framework for Analysis* (Livermore, CA: Lawrence Livermore National Laboratory, October 2000).

8. Elisa Wood, "Microgrid 2017: ComEd CEO Sees Microgrids Emerging as 'Defining Infrastructure,'" Microgrid Knowledge, November 10, 2017, https://microgridknowledge.com/microgrid-2017 -com-ed.

9. Paul Hawken, *Drawdown: The Most Comprehensive Plan Ever Proposed to Reverse Global Warming* (New York: Penguin, 2017).

10. David Roberts, "Utilities Have a Problem: The Public Wants 100% Renewable Energy, and Quick," Vox, September 14, 2018, https://www.vox.com/energy-and-environment/2018/9/14/17853884/ utilities-renewable-energy-100-percent-public-opinion.

11. David Roberts, "The Key to Tackling Climate Change: Electrify Everything," Vox, September 19, 2016, https://www.vox .com/2016/9/19/12938086/electrify-everything.

12. Greg Satell, "How the Energy Revolution Will Transform How We Live and Work," *Forbes*, February 8, 2015, www.forbes.com/ sites/gregsatell/2015/02/08/how-the-energy-revolution -will-transform-how-we-live-and-work/.

13. Gretchen Bakke, *The Grid: The Fraying Wires between Americans and Our Energy Future* (New York: Bloomsbury, 2016). Bakke's best seller was on Bill Gates's book picks list for 2016.

14. "2017 Infrastructure Report Card," American Society of Civil Engineers, 2017, https://www.infrastructurereportcard.org.

15. Michel Callon, Pierre Lascoumes, and Yannick Barthe, *Acting in an Uncertain World: An Essay on Technical Democracy* (Cambridge, MA: MIT Press, 2009).

16. Siskin, *System*.

17. Wikipedia, "Adaptation," Fall 2017. Alas, this definition around self-adapting systems seems self-referential.

18. Donella Meadows, *Thinking in Systems* (White River Junction, VT: Chelsea Green, 2008), 145–165, 86.

19. Siskin, *System*, 30, 3.

20. Ian Bogost, "You Are Already Living Inside a Computer," *Atlantic*, September 14, 2017, https://www.theatlantic.com/technology/archive/2017/09/you-are-already-living-inside-a -computer/539193.

21. "U.S. Microgrids 2017: Market Drivers, Analysis and Forecast," Wood Mackenzie, November 2017, https://www .greentechmedia.com/research/report/us-microgrids-2017.

22. Across this period of writing, 2016–2018, I have steadily monitored feeds from Greentech Media, *Microgrid Knowledge*, Navigant Research, Rocky Mountain Institute, Smart Electric Power Alliance, and more.

23. Ian Hacking, *Historical Ontology* (Cambridge, MA: Harvard University Press, 2002).

24. For reliable standard definitions and measures in microgrids, without a stake in building the results, this project relies on the work of the Lawrence Berkeley National Laboratory, which I visited for a month in fall 2016, while on academic leave and researching this book.

25. Office of Electricity, "The Role of Microgrids in Helping to Advance the Nation's Energy System," US Department of Energy, 2017, https://www.energy.gov/oe/activities/technology-development/ grid-modernization-and-smart-grid/role-microgrids-helping.

26. Wood, "Microgrid 2017."

2   Electrification's Eras

1. Richard Pence, ed., *The Next Greatest Thing* (Washington, DC: National Rural Electric Cooperative Association, 1984), 1.

2. Morris Cooke, "Plan for a Nation-Wide Development of Rural Electrification," in *The Next Greatest Thing*, ed. Richard Pence (Washington, DC: National Rural Electric Cooperative Association, 1984), 61–62.

3. David Cushman Coyle, ed., *Electric Power on the Farm* (Washington, DC: US Rural Electrification Administration, 1936).

4. Thomas P. Hughes, *Networks of Power: Electrification in Western Society, 1880–1930* (Baltimore: Johns Hopkins University Press, 1983), 17, 7, 6.

5. Vijay Vaitheesswaran, *Power to the People: How the Coming Energy Revolution Will Transform an Industry, Change Our Lives, and Maybe Even Save the Planet* (New York: Farrar, Straus and Giroux, 2003), 32.

6. Walt Patterson, *Transforming Electricity: The Coming Generation of Change* (London: Earthscan 1999), 1, 12–13.

7. Gretchen Bakke, *The Grid: The Fraying Wires between Americans and Our Energy Future* (New York: Bloomsbury, 2016).

8. David E. Nye, *America's Assembly Line* (Cambridge, MA: MIT Press. 2013).

9. Tyler Hamilton, *Mad Like Tesla: Underdog Inventors and the Relentless Pursuit of Clean Energy* (Toronto: ECW Press, 2011).

10. Bakke, *The Grid*.

11. Of the many histories on the Columbian Exposition, see Harold Platt, *The Electric City: Energy and the Growth of the Chicago Area, 1880–1930* (Chicago: University of Chicago Press, 1991).

12. David E. Nye, *Electrifying America: Social Meanings of a New Technology, 1880–1940* (Cambridge, MA: MIT Press, 1990), 37–41.

13. "Jeremiah D. Lambert, *The Power Brokers: The Struggle to Shape and Control the Electric Power Industry* (Cambridge, MA: MIT Press, 2015), 10.

14. Bakke has sketched this origin well: "The hydroelectric plant at Niagara Falls was the closing bell on the effervescent, chaotic immensely creative and inventive activity of the previous seventeen years. 1879, the first arc light grid in San Francisco; 1882, the first low-voltage, direct current grid in New York; 1887, the first alternating current grid; 1891, proven long-distance high voltage transmission. And in 1896, the completion of the first large-scale generating station at Niagara Falls, together with the first long-distance transmission wires in constant use, the total adoption of parallel circuits, incandescent lighting, and the equal near total adoption of alternating current. America had her grid." Bakke, *The Grid*, 54.

15. E. T. Whitaker, *A History of the Theories of Aether and Electricity: From the Age of Descartes to the Close of the Nineteenth Century* (London: Longmans, Green, and Co., 1910), 1.

16. Linda Simon, *Dark Light: Electricity and Anxiety from the Telegraph to the X-ray* (Boston: Houghton Mifflin Harcourt, 2004), 3.

17. Simon, *Dark Light*, 4.

18. Simon, *Dark Light*, 169.

19. Nye, *Electrifying America*, 156, 138–184.

20. Nye, *Electrifying America*, 156.

21. Thomas P. Hughes, "How to Think about Technology and Culture," in *Human-Built World: How to Think about Technology and Culture* (Chicago: University of Chicago Press, 2005), 111–132.

22. Nye, *Electrifying America*, 138.

23. Joel Tarr and Gabriel Dupuy, eds., *Technology and the Rise of the Networked City in Europe and America* (Philadelphia: Temple University Press, 1988).

24. Ithiel de Sola Pool, ed., *The Social Impact of the Telephone* (Cambridge, MA: MIT Press, 1977).

25. Lewis Mumford, *Sticks and Stones: A Study of American Architecture and Civilization* (New York: Boni, 1924), 163, 164.

26. To me, as a resident of metro Detroit, these are local legends. On all this as an effect of electrification, Nye has researched and told it well in *America's Assembly Line*.

27. Reyner Banham, *Theory and Design in the First Machine Age* (Cambridge, MA: MIT Press, 1960), 10.

28. Filippo Tomasso Marinetti, cited in Banham, *Theory and Design in the First Machine Age*, 125.

29. Talbot Faulkner Hamlin, *The American Spirit in Architecture* (New Haven, CT: Yale University Press, 1926), 258.

30. Banham, *Theory and Design in the First Machine Age*, 11.

31. Elizabeth Shove, *Comfort, Cleanliness, Convenience: The Social Organization of Normality* (Oxford: Berg, 2003), 117–158.

32. Ronald Tobey, *Technology as Freedom: The New Deal and the Electrical Modernization of the American Home* (Berkeley: University of California Press, 1996), 155–166.

33. "The New Deal shifted the majority of American families to an asset strategy for economic security through state-enframed home ownership of electrically modern dwellings." Tobey, *Technology as Freedom*, 209.

34. Tobey, *Technology as Freedom*, 69–71. Aside from lights, and the flat iron, which 80 percent of electrified and 60 percent of all homes owned, no other electrical appliance was held by a majority, with the vacuum cleaner, second most, owned by just 20 percent of all homes, and the clothes washer by 10 percent.

35. Platt, *The Electric City*, 269.

36. David E. Nye, *When the Lights Went Out: A History of Blackouts in America* (Cambridge, MA: MIT Press, 2010), 144.

37. Lambert, *The Power Brokers*, 19, citing Samuel Insull, *Central Station Electric Service*, ed. William Eugene Kiely (Chicago, private printing, 1915), 116–117.

38. Bakke, *The Grid*, 66.

39. Lambert, *The Power Brokers*, 15

40. Lambert, *The Power Brokers*, 49.

41. Lambert, *The Power Brokers*, 49, citing Platt, *The Electric City*, 276.

42. Institute for Energy Research, July 2014, https://www.instituteforenergyresearch.org.

43. Gifford Pinchot, cited in Morris Cooke, "Plan for a Nation-Wide Development of Rural Electrification," cited in Pence, *The Next Greatest Thing*, 61–62.

44. Hughes (*Networks of Power*, 370) provides an oft-cited summary:

    1. Economies of scale, in steam turbine generation
    2. Locating these giant power plants near population centers
    3. High-voltage transmission to load centers
    4. Cultivating mass consumption especially via rate differentials
    5. Interconnecting power plants to optimize their different characteristics
    6. Interconnection of loads to aid diversification and balancing
    7. Centralizing control
    8. Forecasting loads
    9. Lowering peak capacity and scheduling maintenance via interconnection
    10. Accepting government regulation to establish natural monopoly
    11. Achieving consistent return on investment so as to obtain affordable capital

45. Hughes, *Networks of Power*, 1.

46. Bakke, *The Grid*, xviii.

3 Smart Green Blues

1. Lester Brown, *The Great Transition: Shifting from Fossil Fuels to Solar and Wind Energy* (New York: W. W. Norton, 2015), 17, 150. Here he continues, "Our relation with the natural world will change from one where we are in conflict with nature to one where we are in sync with it."

2. Brown, *The Great Transition*, 141.

3. Gretchen Bakke, *The Grid: The Fraying Wires between Americans and Our Energy Future* (New York: Bloomsbury, 2016).

4. Brown, *The Great Transition*, 76.

5. Trevor Pinch and Wiebe Bijker, "The Social Construction of Facts and Artifacts: Or How the Sociology of Science and the Sociology of Technology Might Help Each Other," in *The Social Construction of Technological Systems*, ed. Wiebe Bijker, Trevor Pinch, and Thomas P. Hughes (Cambridge, MA: MIT Press, 2012), 11–44. "By this we mean that there is not only flexibility in how people think of or interpret artifacts but also that there is flexibility in how artifacts are designed. There is not just one possible way or one best way of designing an artifact." Pinch and Bijker, "The Social Construction of Facts and Artifacts," 34.

6. Pinch and Bijker, "The Social Construction of Facts and Artifacts," 20–21.

7. Pinch and Bijker, "The Social Construction of Facts and Artifacts," 20–21.

8. Richard Hirsh, *Power Loss: The Origins of Deregulation and Restructuring in the American Electric Utility System* (Cambridge, MA: MIT Press, 1999), 11–31, 2, 68.

9. Jeremiah D. Lambert, *The Power Brokers: The Struggle to Shape and Control the Electric Power Industry* (Cambridge, MA: MIT Press 2015), 100–130.

10. Lambert, *The Power Brokers*, 262.

11. Paul Joskow and Richard Schmalensee, *Markets for Power: An Analysis of Electric Utility Deregulation* (Cambridge, MA: MIT Press, 1983).

12. Joskow and Schmalensee, *Markets for Power*.

13. The histories of Enron are many. See, for example, Lambert, *The Power Brokers*; Vijay Vaitheesswaran, *Power to the People: How the Coming Energy Revolution Will Transform an Industry, Change Our Lives, and Maybe Even Save the Planet* (New York: Farrar, Straus and Giroux, 2003).

14. David E. Nye, *When the Lights Went Out: A History of Blackouts in America* (Cambridge, MA: MIT Press, 2010), 33.

15. Mark Chediak, "U.S. Power Demand Flatlined Years Ago, and It's Hurting Utilities," Bloomberg, April 25, 2017, https://www.bloomberg.com/news/articles/2017-04-25/u-s-power-demand-flatlined-years-ago-and-it-s-hurting-utilities.

16. Amory Lovins, *Reinventing Fire: Bold Business Solutions for the New Energy Era* (White River Junction, VT: Chelsea Green, 2011).

17. Mason Wilrich, *Modernizing America's Electricity Infrastructure* (Cambridge, MA: MIT Press, 2017).

18. Peter Fox-Penner, *Smart Power: Climate Change, the Smart Grid, and the Future of Electric Utilities* (Washington, DC: Island Press, 2014), xiii.

19. Jennie Stephens, Elizabeth Wilson, and Tarla Rai Peterson, *Smart Grid (R)Evolution: Electric Power Struggles* (New York: Cambridge University Press, 2015).

20. Author interview with Gretchen Bakke, July 2017.

21. Comment from an anonymous reader of this book's manuscript.

22. John Farrell, "Reverse Power Flow: How Solar+Batteries Shift Electric Grid Decision Making from Utilities to Consumers," Institute for Local Self-Reliance, July 2018, https://ilsr.org/solar-plus-storage.

23. Stephens, Wilson, and Peterson, *Smart Grid (R)evolution*, 7–8, 186.

24. Rebecca Slayton, "Efficient, Secure Green: Digital Utopianism and the Challenge of Making the Electrical Grid 'Smart,'" *Information and Culture* 48, no. 4 (2013): 450.

25. Electric Power Research Institute, 2017, http://smartgrid.epri.com.

26. John Doerr, "Salvation (and Profit) in Greentech," TED, 2007, https://www.ted.com/talks/john_doerr_sees_salvation_and_profit_in_greentech.

27. Venkat Pothamsetty and Saadat Malik, "Smart Grid Leveraging Intelligent Communications to Transform the Power Infrastructure," Cisco white paper, February 2009, 1, 4, 9, 11.

28. Cisco, "Digital Utilities" (2018), https://www.cisco.com/c/en/us/solutions/industries/energy/external-utilities-smart-grid.html.

29. Mike Munsell, "GTM's Grid Edge Executive Council Now Counts 100 Member Companies," GTM, January 28, 2016, https://www.greentechmedia.com/articles/read/gtms-grid-edge-executive-council-now-counts-100-member-companies.

30. Nick Srnicek, *Platform Capitalism* (Cambridge, UK: Polity Press, 2016).

31. Stuart Bennett, *A History of Control Engineering, 1930–1955* (Washington, DC: Institution of Engineering and Technology, 1993).

32. Stuart Bennett, *A History of Control Engineering, 1800–1930* (Washington, DC: Institution of Engineering and Technology, 1979).

33. Gene Smith, "Power Industry Adds Computers," *New York Times*, October 8, 1964), 63. The *New York Times* ran several articles by Smith on power automation.

34. Slayton, "Efficient, Secure Green," 458.

35. Fox-Penner, *Smart Power*, 25–29.

36. Adam Greenfield, *Against the Smart City* (London: Verso, 2013), kindle segments 4 and 5 respectively.

37. Carolyn Marvin, *When Old Technologies Were New: Thinking about Electric Communication in the Late Nineteenth Century* (New York: Oxford University Press, 1988), 4.

38. Elizabeth Shove, "Infrastructures and Practices: Networks beyond the City," in *Beyond the Networked City*, ed. Olivier Coutard and Jonathan Rutherford (London: Routledge, 2016), 242–257.

39. Evgeni Morozev, *To Save Everything, Click Here* (New York: Public Affairs, 2013).

40. This refers to a literature search of Social History of Technology and the Society for the Social Studies of Science papers, 2007–2015, conducted in fall 2016.

41. Author interview with Rebecca Slayton, May 2016.

42. Author interview with Phoebe Sengers, July 2016.

43. A Google search for "green grid blues" yielded zero results in spring 2017; likewise for "smart green blues" in fall 2017.

4   Microgrid Institutions

1. Elisa Wood, "How 2017 Made the Argument for Microgrids," Microgrid Knowledge, January 2, 2018, https://microgridknowledge.com/microgrids-2017.

2. Umair Irfan, "Puerto Rico's Blackout, the Largest in American History, Explained," Vox, May 8, 2018, https://www.vox.com/2018/2/8/16986408/puerto-rico-blackout-power-hurricane.

3. Naomi Klein, "The Battle for Paradise," Intercept, March 20, 2018, https://theintercept.com/2018/03/20/puerto-rico-hurricane-maria-recovery.

4. Deborah Acosta and Frances Robles, "Puerto Ricans Ask: When Will the Lights Come Back On?," *New York Times*, October 19, 2017, https://www.nytimes.com/2017/10/19/us/puerto-rico-electricity-power.html.

5. Peter Fox-Penner, "Why Solar 'Microgrids' Are Not a Cure-all for Puerto Rico's Power Woes," Conversation, November 8, 2017, https://theconversation.com/why-solar-microgrids-are-not-a-cure-all-for-puerto-ricos-power-woes-86437.

6. Fox-Penner, "Why Solar 'Microgrids' Are Not a Cure-all."

7. Resilient Power Puerto Rico, "Our Mission," 2017, https://resilientpowerpr.org/about-us.

8. Mike Scott, "Microgrids Hold the Key to Providing Power for All," *Forbes*, July 19, 2018, https://www.forbes.com/sites/mikescott/2018/07/19/microgrids-hold-the-key-to-providing-power-for-all/#59bf8c2d6472.

9. Julia Pyper, "Large Corporations Are Driving America's Renewable Energy Boom. And They're Just Getting Started," Greentech Media, January 10, 2017, https://www.greentechmedia.com/articles/read/large-corporations-are-driving-americas-renewable-energy-boom.

10. Peter Asmus, "Here Come the 'Locavolts!,'" *Renewable Energy Resilience*, July 20, 2008, www.peterasmus.com/journal/2008/7/20/here-come-the-locavolts.html. See also Jennie Stephens, Elizabeth Wilson, and Tarla Rai Peterson, *Smart Grid (R)Evolution: Electric Power Struggles* (New York: Cambridge University Press, 2015), 147.

11. Denise Fairchild and Al Weinrub, *Energy Democracy: Advancing Equity in Clean Energy Solutions* (Washington, DC: Island Press, 2017), 1–20.

12. Dan T. Ton and Merrill A. Smith, "The U.S. Department of Energy's Microgrid Initiative," *Electricity Journal* 25, no. 8 (October 2012): 84.

13. Matt Grimley and John Farrell, "Mighty Microgrids," Institute for Local Self-Reliance, March 3, 2016, https://ilsr.org/report-mighty-microgrids.

14. Robert H. Lasseter, Abbas Akhil, Chris Marnay, John Stephens, Jeff Dagle, Ross Guttromson, A. Sakis Meliopoulous, Robert Yinger, and Joe Eto, "The CERTS MicroGrid Concept," Transmission Reliability Program, US Department of Energy, April 2002, 30.

15. LexisNexis word search of "microgrid," by decade, from 1950 to 2000.

16. See, for instance, Robert H. Lasseter and Paolo Piagi, "Microgrid: A Conceptual Solution," *IEEE Annual Power Electronics Specialists Conference* 6 (July 2004): 4285–4290.

17. Anya Kamenetz, "Why the Microgrid Could Be the Answer to Our Energy Crisis," *Fast Company*, July 1, 2009, https://www.fastcompany.com/1297936/why-microgrid-could-be-answer-our-energy-crisis.

18. Elisa Wood, ed., "Think Microgrid: A Discussion Guide for Policymakers, Regulators and End Users," International District Energy Association, May 2014, https://blog.se.com/wp-content/uploads/2014/06/Think-Microgrid-Special-Report.pdf.

19. *Economist*, "Grid Unlocked: American Utilities Mimic the Tech Industry to Make Systems More Resilient," October 17, 2014, http://www.economist.com/news/business/21625885-american-utilities-mimic-tech-industry-make-systems-more-resilient-grid-unlocked.

20. "Community Microgrid Initiative," Clean Coalition, 2019, https://clean-coalition.org/community-microgrid-initiative.

21. Grimley and Farrell, "Mighty Microgrids," 2.

22. Navigant Research, "Market Data: Microgrids," Summer 2014, https://www.navigantresearch.com/research/market-data-microgrids.

23. Peter Asmus, "Microgrids: Pie-in-the-Sky Dreams versus On-the-Ground Realities," Navigant Research, March 4, 2016, https://www.navigantresearch.com/news-and-views/microgrids-pie-in-the-sky-dreams-versus-on-the-ground-realities.

24. Navigant Research, "Market Data: Energy Storage for Microgrids," Fall 2017, https://www.navigantresearch.com/reports/market-data-energy-storage-for-microgrids; Navigant Research, "Building-to-Grid Integration," Fall 2017, https://www.navigantresearch.com/reports/building-to-grid-integration; Navigant Research, "Navigant Research Leaderboard: Smart City Suppliers," Fall 2017, https://www.navigantresearch.com/reports/navigant-research-leaderboard-smart-city-suppliers.

25. Author interview with Bruce Nordman, June 2016.

26. Bruce Nordman and Ken Christensen, "DC Local Power Distribution with Microgrids and Nanogrids," n.d., https://eta-intranet.lbl.gov/sites/default/files/icdcm2015nordmanLPD.pdf.

27. Chris Marnay, Bruce Nordman, and Judy Lai, "Future Roles of Milli-, Micro-, and Nano-Grids," Microgrids at Berkeley Lab, 2011, https://building-microgrid.lbl.gov/publications/future-roles-milli-micro-and-nano.

28. Elisa Wood, "Solar Is Good. Solar Microgrids Are Better," Microgrid Knowledge, August 8, 2106, https://microgridknowledge.com/solar-microgrids-are-better.

29. Gretchen Bakke, *The Grid: The Fraying Wires between Americans and Our Energy Future* (New York: Bloomsbury, 2016), 219–254.

30. Author interview with Jim Saber, May 2017.

31. That is just what this project has done, alongside its many other methods, including library research, long-term historic reading, mostly academic interviews, and a short residence at the Berkeley Lab. Hence the many journalistic items in these endnotes.

32. David Roberts and Alvin Chang, "Meet the Microgrid, the Technology Poised to Transform Electricity," Vox, May 24, 2018, https://www.vox.com/energy-and-environment/2017/12/15/16714146/greener-more-reliable-more-resilient-grid-microgrids.

33. David Roberts, "The Key to Tackling Climate Change: Electrify Everything," Vox, September 19, 2016, https://www.vox .com/2016/9/19/12938086/electrify-everything.

34. "Microgrid Knowledge Survey 2016," Microgrid Knowledge, http://microgridknowledge.com/white-paper/microgrid-knowledge -survey-2016.

35. Kevin Normandeau, "Microgrid 2018 Opens with the Message: Microgrids for the Greater Good," Microgrid Knowledge, May 7, 2018, https://microgridknowledge.com/microgrids-greater-good.

36. Christopher Villarreal, David Erickson, and Marzia Zafar, "Microgrids: A Regulatory Perspective," California Public Utility Commission, April 14, 2014, 25, 3.

37. "Hudson Yards: Sustainability and Resiliency," Hudson Yards, 2016, http://www.hudsonyardsoffices.com/office/10-hudson-yards/ availabilities/building/sustainability-and-resiliency.

38. Jeffrey Spivak, "A Developing Front in Resilience: Electricity Microgrids," Urban Land, February 17, 2015, https://urbanland.uli .org/sustainability/developing-front-resilience-electricity-microgrids.

39. Diane Cardwell, "Solar Experiment Lets Neighbors Trade Energy among Themselves," *New York Times*, March 13, 2017, https://www.nytimes.com/2017/03/13/business/energy-environment/ brooklyn-solar-grid-energy-trading.html.

40. "The Future of Energy Is Local," Brooklyn Microgrid, http:// brooklynmicrogrid.com.

41. "Powering a New Generation of Community Energy," New York State Energy Research and Development Authority, https:// www.nyserda.ny.gov/All-Programs/Programs/NY-Prize.

42. Pecan Street, www.pecanstreet.org.

43. Stephens, Wilson, and Peterson, *Smart Grid (R)Evolution*, 28–31.

44. Stephens, Wilson, and Peterson, *Smart Grid (R)Evolution,* 150.

45. Author interview with Jennie Stephens, June 2016.

46. Mags Harries and Lajos Heder, SunFlowers installation, 2009, http://muellersunflowers.powerdash.com/aboutus.

47. United Nations Environmental Programme, "District Energy in Cities: Unlocking the Potential of Energy Efficiency and Renewable Energy," 2014, 28.

48. Morgan Kelly, "Two Years after Hurricane Sandy, Recognition of Princeton's Microgrid Still Surges," Princeton University, October 23, 2014, https://www.princeton.edu/main/news/archive/S41/40/10C78/index.xml?section=featured.

49. Edward Borer, "Resilient Community Microgrids: Concepts, Operations and Maintenance," International District Energy Association,February 2015, http://www.microgridresources.com/HigherLogic/System/D ownloadDocumentFile.ashx?DocumentFileKey=38b01592-a6a4-5446-218d-73b998943030.

50. Daniel Kammen, "The Ecoblock Project," Renewable and Appropriate Energy Laboratory, University of California at Berkeley. https://rael.berkeley.edu/project/the-eco-block-project. See also "Sustainable Design of Communities," *Scientific American*, June 26, 2017, https://rael.berkeley.edu/2017/06/sustainable-design-of-communities-in-scientific-american.

51. Elisa Wood, "Pittsburgh Plans Grid of Microgrids; Wins Smart City Money," Microgrid Knowledge, October 17, 2016, https://microgridknowledge.com/grid-of-microgrids-pittsburgh.

52. John Farrell, "Energy Democracy in the 51st State," Institute for Local Self-Reliance, February 2015.

53. Mary Douglas, *How Institutions Think* (Syracuse, NY: Syracuse University Press, 1986), 55.

54. Kevin Lynch, *The Image of the City* (Cambridge, MA: MIT Press, 1960), 49–52.

55. Lynch, *The Image of the City*, 46–90.

56. Lucy Bullivant, ed., *4D Hyperlocal: A Cultural Toolkit for the Open-Source City* (Hoboken, NJ: Wiley, 2017).

5   Architecture's Grid Edge

1. This generalization about buildings comprises residential, commercial, information-technological, and industrial uses, which are quite distinct in their usage issues and patterns. This project research included a visit to the Lawrence Berkeley National Laboratory, which measures such patterns at a vast societal scale, and whose data reports (for instance, the famous Sankey diagrams of large-scale energy flow) are the most widely cited of their kind.

2. Word counts on Google searches, March 2017.

3. Digital fabrication, design for assembly, and material systems have become the most prominent research focus of my own institution, Taubman College of Architecture and Urban Planning at the University of Michigan, and this focus provides at least some basis for investigating architecture's grid edge as a material system design opportunity.

4. Bruce Nordman, "Beyond the Smart Grid: Building Networks," Lawrence Berkeley National Laboratory, 2010, https://eta-intranet .lbl.gov/sites/default/files/beyond.pdf. Nordman's many clear writings and presentations are an especially prominent resource from Lawrence Berkeley National Laboratory.

5. On building energy, I am grateful for guidance from several kind experts in this aspect of my home discipline. See my acknowledgments. To me the most useful spot-check has been the Architecture and Energy conference held at Penn in 2012, and published as William Braham, and Daniel Willis, eds, *Architecture and Energy: Performance and Style* (New York: Routledge, 2013).

6. This has been the emphasis of my previous writings, the most enduring of which has been Malcolm McCullough, *Digital Ground: Architecture, Pervasive Computing, and Environmental Knowing* (Cambridge, MA: MIT Press, 2005).

7. For a representative sample of contemporary technologies and agendas, see the working papers at BuildingIQ, https://buildingiq .com/resources.

8. Kevin Lucas, "So What Exactly Is This 'Grid Edge' Thing, Anyway?," Alliance to Save Energy, June 20, 2016, https://www .ase.org/blog/so-what-exactly-grid-edge-thing-anyway. "Grid Edge hardware is the physical material you can touch: solar panels, advanced metering infrastructure, smart inverters, energy storage systems, smart thermostats, smart appliances and building controls."

9. "Why Are Buildings Behaving Badly?" Grid Edge Limited, 2017, http://www.gridedge.co.uk/buildings-behaving-badly-1.

10. "World Green Building Trends 2018: SmartMarket Report," World Green Building Council, November 13, 2018, 18–19.

11. Tom Madden, "The Greenest Building in the World: The Edge," Living Map, August 3, 2017, https://www.livingmap.com/ smart-building/the-edge.

12. "The Edge: Amsterdam, The Netherlands," PLP Architecture, 2016, http://www.plparchitecture.com/the-edge.html.

13. Tom Randall, "The Smartest Building in the World: Inside the Connected Future of Architecture," Bloomberg, September 23, 2015, https://www.bloomberg.com/features/2015-the-edge-the-worlds-greenest-building.

14. Jane Wakefield, "Will Tomorrow's Smart Office Be a Saviour or a Spy?," BBC News, April 6, 2016, http://www.bbc.com/news/technology-35696521.

15. "Elon Musk Explains Why the Tesla-SolarCity Merger Is All about Microgrids," Microgrid Media, June 27, 2016, http://microgridmedia.com/elon-musk-explains-why-the-tesla-solarcity-merger-is-all-about-microgrids.

16. See Association for Computer Aided Design in Architecture, acadia.org.

17. William Mitchell, *Me++: The Cyborg Self and the Networked City* (Cambridge, MA: MIT Press, 2003), 7–17.

18. Mitchell, *Me++*, 7, 10

19. Mitchell, *Me++*, 10.

20. William Mitchell, *Placing Words: Symbols, Space, and the City* (Cambridge, MA: MIT Press, 2005), 9.

21. As this gross generalization has been unpacked in some of my previous work, please excuse such a short summary here. See, for instance, Malcolm McCullough, *Ambient Commons: Attention in the Age of Embodied Information* (Cambridge, MA: MIT Press, 2013), 91–108.

22. Mitchell, *Placing Words*, 4.

23. Most first citations of the extended mind are to the body of work by Andy Clark, including a later book by that very name. For the standard citation, see Andy Clark and David J. Chalmers, "The Extended Mind," *Analysis* 58, no. 1 (January 1998): 7–19.

24. Ian Bogost, *Alien Phenomenology, or, What It's Like to Be a Thing* (Minneapolis: University of Minnesota Press, 2012).

25. Elizabeth Shove, "Infrastructures and Practices: Networks beyond the City," in *Beyond the Networked City*, ed. Olivier Coutard and Jonathan Rutherford (London: Routledge, 2016), 246.

26. "H-E-B at Mueller," American Institute of Architects, 2016, http://www.aiatopten.org/node/489.

27. Lake/Flato, 2016, http://www.lakeflato.com/urban-development/h-e-b-mueller.

28. Author interview with Stephen Selkowitz, Lawrence Berkeley National Laboratory, September 2016.

29. "A Day in the Light," *Metropolis*, May 2004, http://www .metropolismag.com/uncategorized/a-day-in-the-light.

30. "Daylighting the New York Times Building," Lawrence Berkeley National Laboratory, https://facades.lbl.gov/newyorktimes/ nyt_arch-approach.html.

31. Author interview with Stephen Selkowitz, Lawrence Berkeley National Laboratory, September 2016.

6   Situated Interactions

1. Howard Gardner and Katie Davis, *The App Generation: How Today's Youth Navigate Identity, Intimacy, and Imagination in a Digital World* (New Haven, CT: Yale University Press, 2013).

2. Rebecca Solnit, "Diary: In the Day of the Postman," *London Review of Books* 35, no. 16 (August 29, 2013): 32–33.

3. Clifford Siskin, *System: The Shaping of Modern Knowledge* (Cambridge, MA: MIT Press, 2016).

4. At this writing, the annual ACM Tangible and Embedded Interaction conference, now in its twelfth year, has adopted the currently popular theme of "postdigital."

5. Author interview with Hal Wilhite, September 2016.

6. Author interview with Harry Giles, May 2016.

7. Grégoire Wallenborn and Harold Wilhite, "Rethinking Embodied Knowledge and Household Consumption," *Energy Research and Social Science* 1 (2014): 56–64.

8. Douglas Rushkoff, *Present Shock: When Everything Happens Now* (New York: Penguin, 2013).

9. In 2007, *Wired* magazine editors Gary Wolf and Kevin Kelly started a "life-logging" movement. Today, many resources link to Wolf's blog, *Quantified Self*. http://quantifiedself.com.

10. Gardner and Davis, *The App Generation*.

11. Thomas Tierney, *The Value of Convenience* (Albany: SUNY Press, 1993), 39–40.

12. Lewis Mumford, *Technics and Civilization* (New York: Harcourt, Brace, 1936).

13. See especially Tim Wu, *The Attention Merchants: The Epic Scramble to Get Inside Our Heads* (New York: Alfred A. Knopf, 2016).

14. Elizabeth Shove, *Comfort, Cleanliness, Convenience: The Social Organization of Normalcy* (Oxford, UK: Berg, 2003), 184, 171–183, 196.

15. Jennie Stephens, Elizabeth Wilson, and Tarla Rai Peterson, *Smart Grid (R)Evolution: Electric Power Struggles* (New York: Cambridge University Press, 2015), 186.

16. Dietmar Offenhuber, "Indexical Visualization: Traces, Symptoms, Evidence" (working paper, spring 2018).

17. Author interview with Rick Diamond, June 2016.

18. For the classic argument that comfort is best understood as a pursuit and not a state, see Lisa Heschong, *Thermal Delight in Architecture* (Cambridge, MA: MIT Press, 1978).

19. Author interview with Chris Payne, September 2016.

20. Richard Coyne, *The Tuning of Place: Sociable Spaces and Pervasive Digital Media* (Cambridge, MA: MIT Press, 2010), 19, xxvii.

21. Yolande Strengers, *Smart Energy Technologies in Everyday Life: Smart Utopia?* (London: Palgrave Macmillan, 2013), 2–10.

22. Strengers, *Smart Energy Technologies in Everyday Life*, 8, 135, 151–152.

23. James Pierce and Eric Paulos, "Materializing Energy," in *Proceedings of the 8th ACM Conference on Designing Interactive System* (New York: ACM, 2010), 113–122.

24. Although this project began from past work in ambient interface, it has alas witnessed a deceleration of tangibly interactive arts and installations. Built space has often become more responsive, but seldom with nonscreen engagement.

25. Wallenborn and Wilhite, "Rethinking Embodied Knowledge and Household Consumption," 56.

26. Matt Crawford, *The World beyond Your Head: On Becoming an Individual in an Age of Distraction* (New York: Farrar, Straus and Giroux, 2015).

27. At least among my own previous works, there is a common thread on presence and practice through flowing engagement with situated technologies. This is not the higher meaning of the perennial philosophy per Aldous Huxley, but just a persistence about craft.

28. Shove, *Comfort, Cleanliness, Convenience*, 196.

29. This has been a theme on attention in my previous book; see Malcolm McCullough, *Ambient Commons: Attention in the Age of Embodied Information* (Cambridge, MA: MIT Press, 2013). See also Brian Bruya, *Effortless Attention: A New Perspective in the Cognitive Science of Attention and Action* (Cambridge, MA: MIT Press, 2010), 3–29.

7   To Island

1. Rebecca Solnit, *A Paradise Built in Hell: The Extraordinary Communities That Arise in Disaster* (New York: Penguin, 2009).

2. UN Office for Disaster Risk Reduction, "How to Make Cities More Resilient: A Handbook for Local Government Leaders" (Geneva: United Nations, 2012), 63.

3. "Definitions of Community Resilience: An Analysis," Community and Regional Resilience Institute, 2013, http://www.resilientus.org/wp-content/uploads/2013/08/definitions-of-community-resilience.pdf.

4. "Developing Urban Resilience," Urban Land Institute, 2017, https://developingresilience.uli.org.

5. Daniel Lerch, "Six Foundations for Building Community Resilience," Post Carbon Institute, November 18, 2015, 4.

6. "Definitions of Community Resilience: An Analysis."

7. Kresge Foundation, "Bounce Forward: Urban Resilience in the Era of Climate Change" (Washington, DC: Island Press, 2015).

8. "Community Microgrid Initiaitive," Clean Coalition, 2019, https://clean-coalition.org/community-microgrid-initiative.

9. Lerch, "Six Foundations for Building Community Resilience," 15.

10. Author interview with Bruce Nordman, September 2016.

11. Andrew Zolli and Anne Marie Healy, *Resilience: Why Things Bounce Back* (New York: Simon and Schuster, 2012), 25–28, 115, 92–100.

12. Lance H. Gunderson and C. .S. Holling, eds., *Panarchy: Understanding Transformations in Human and Natural Systems* (Washington, DC: Island Press, 2002).

13. Clifford Siskin, *System: The Shaping of Modern Knowledge* (Cambridge, MA: MIT Press, 2016), 3–21.

14. Daniel Daou and Pablo Pérez-Ramos, eds., *New Geographies, 8: Island* (Cambridge, MA: Harvard Graduate School of Design, 2016), 7. The existence of this intriguing volume, which is devoted more to poetics, has helped me recast some larger argument about clusters here into metaphors of the archipelago, which help explain the resonance of the microgrid meme.

15. Immanuel Kant, *Critique of Pure Reason* (1781), trans. Norman Kemp Smith (London: Macmillan 1964), 257, cited in Robin Mackay, "Philosophers' Islands," in *New Geographies, 8: Island*, ed. Daniel Daou and Pablo Pérez-Ramos (Cambridge, MA: Harvard Graduate School of Design, 2016), 61.

16. Marc Shell, *Islandology: Geography, Rhetoric, Politics* (Stanford, CA: Stanford University Press, 2014), 13.

17. Sonia Glavaski, "Controlling the Grid Edge," ARPA-E, September 2018, https://arpa-e.energy.gov/sites/default/files/ Glavaski_FastPitch.pdf.

18. David Roberts, "Clean Energy Technologies Threaten to Overwhelm the Grid: Here's How It Can Adapt," Vox, December 3, 2018, https://www.vox.com/energy-and-environment/2018/11/30 /17868620/renewable-energy-power-grid-architecture.

19. Gretchen Bakke, *The Grid: The Fraying Wires between Americans and Our Energy Future* (New York: Bloomsbury, 2016), xviii.

20. E. R. Morgan, S. Valentine, C. A. Blomberg, E. R. Limpaecher, and E. V. Dydek, "Boston Community Energy Study: Zonal Analysis for Urban Microgrids," MIT Lincoln Laboratory, March 2016, https://apps.dtic.mil/dtic/tr/fulltext/u2/1033735.pdf.

21. Goncalo Cardoso, "Distributed Energy Resources: Customer Adoption Model (DER-CAM)," Lawrence Berkeley National Laboratory, May 12, 2016, https://building-microgrid.lbl.gov/ projects/der-cam.

22. Adam Greenfield, *Against the Smart City* (London: Verso, 2013).

23. Oswald Matias Ungers, Rem Koolhaas, Peter Riemann, Hans Kollhoff, and Arthur Ovaska, "The City in the City—Berlin: A Green Archipelago," *Lotus* 19 (1978): 82–97.

24. Oswald Matias Ungers, Rem Koolhaas, Peter Riemann, Hans Kollhoff, and Arthur Ovaska, *The City in the City—Berlin: a Green Archipelago*, ed. Florian Hertweck and Sébastien Marot (Baden, Ger.: Lars Mueller Publishers, 2013).

25. Pier Vittorio Aureli, *The Possibility of an Absolute Architecture* (Cambridge, MA: MIT Press, 2011), 18, 9.

26. Aureli, *The Possibility of an Absolute Architecture*, 46.

27. Aureli, *The Possibility of an Absolute Architecture*, 46.

28. "Navigant Research Leaderboard: Smart City Suppliers," Navigant Research, Fall 2017, https://www.navigantresearch.com/reports/navigant-research-leaderboard-smart-city-suppliers.

29. Drew Hemment and Anthony Townsend, eds., *Smart Citizens* (Manchester, UK: Future Everything, 2013), 5, 9.

30. Christopher Alexander, "A City Is Not a Tree," *Architectural Forum* 122, no. 1 (April 1965): 58–62; Shannon Mattern, "A City Is Not a Computer," *Places Journal* (February 2017), https://placesjournal.org/article/a-city-is-not-a-computer.

31. Mattern, "A City Is Not a Computer."

32. Bruce Sterling, "Stop Saying 'Smart Cities': Digital Stardust Won't Magically Make Future Cities More Affordable or Resilient," *Atlantic*, February 12, 2018, https://www.theatlantic.com/technology/archive/2018/02/stupid-cities/553052.

33. Since in many ways intermittency is where the project began, I was glad to hear a highly respected sociologist raise this too. Author interview with Elizabeth Shove, July 2017.

34. Antoine Picon, "The Limits of Intelligence on the Challenges Faced by Smart Cities," in *New Geographies, 7: Geographies of Information*, ed. Ali Fard and Taraneh Meshkani (Cambridge, MA: Harvard Graduate School of Design, 2015), 77–83.

35. Emily St. John Mandel, *Station Eleven* (New York: Alfred A. Knopf, 2014), 311.

INDEX